T0205420

Springer Theses

Recognizing Outstanding Ph.D. Research

Aims and Scope

The series "Springer Theses" brings together a selection of the very best Ph.D. theses from around the world and across the physical sciences. Nominated and endorsed by two recognized specialists, each published volume has been selected for its scientific excellence and the high impact of its contents for the pertinent field of research. For greater accessibility to non-specialists, the published versions include an extended introduction, as well as a foreword by the student's supervisor explaining the special relevance of the work for the field. As a whole, the series will provide a valuable resource both for newcomers to the research fields described, and for other scientists seeking detailed background information on special questions. Finally, it provides an accredited documentation of the valuable contributions made by today's younger generation of scientists.

Theses are accepted into the series by invited nomination only and must fulfill all of the following criteria

- They must be written in good English.
- The topic should fall within the confines of Chemistry, Physics, Earth Sciences, Engineering and related interdisciplinary fields such as Materials, Nanoscience, Chemical Engineering, Complex Systems and Biophysics.
- The work reported in the thesis must represent a significant scientific advance.
- If the thesis includes previously published material, permission to reproduce this must be gained from the respective copyright holder.
- They must have been examined and passed during the 12 months prior to nomination.
- Each thesis should include a foreword by the supervisor outlining the significance of its content.
- The theses should have a clearly defined structure including an introduction accessible to scientists not expert in that particular field.

More information about this series at http://www.springer.com/series/8790

Lydia Audrey Beresford

Searches for Dijet Resonances

Using $\sqrt{s} = 13$ TeV Proton–Proton Collision Data Recorded by the ATLAS Detector at the Large Hadron Collider

Doctoral Thesis accepted by
the University of Oxford, UK

Springer

Author
Dr. Lydia Audrey Beresford
University of Oxford
Oxford, UK

Supervisors
Prof. B. Todd Huffman
University of Oxford
Oxford, UK

Prof. Çiğdem İşsever
University of Oxford
Oxford, UK

ISSN 2190-5053 ISSN 2190-5061 (electronic)
Springer Theses
ISBN 978-3-030-07365-7 ISBN 978-3-319-97520-7 (eBook)
https://doi.org/10.1007/978-3-319-97520-7

Softcover re-print of the Hardcover 1st edition 2018

This Springer imprint is published by the registered company Springer Nature Switzerland AG
The registered company address is: Gewerbestrasse 11, 6330 Cham, Switzerland

Supervisor's Foreword

In 2012, the field of physics reached a major crossroad. The Higgs boson was discovered using the Large Hadron Collider (LHC) at the European Organization for Nuclear Research (CERN). This particle marked the last significant prediction of a Standard Model of particle physics which had been developed in more or less its final form by the end of the 1970s. This Standard Model as a theoretical construct was highly compelling. It predicted the existence of a host of particles and the strength of the forces between them. It has been remarked that one would lose every bet if one bet against the Standard Model in any area where it claims to make a prediction. This remark has indeed proven to be true.

Despite the success of the Standard Model, other experimental sources indicate the presence of new phenomena about which the Standard Model is silent. The Standard Model provides an excellent description of the interactions of normal matter within the context of the weak and strong nuclear forces and subject to the electromagnetic force, but normal matter and energy only make up 5% of the total energy in our Universe. It is completely silent regarding dark matter, which makes up another 26%, and also regarding dark energy which seems to account for the balance and is causing the Universe's expansion to accelerate.

Up until 2012 and certainly since then, it has become clear that our theoretical model needs to be extended. We have constructed a machine, the Large Hadron Collider, which is now on a mission of exploration that is not directed by any one theoretical construct. We have embarked upon an unrestricted search for evidence which would aid physicists in, perhaps, developing a new, more complete construct that might reveal even deeper secrets into the nature of matter, energy and space-time. At the LHC, energy in the form of protons travelling near to the speed of light is used to try to create new states of matter.

Dr. Lydia Beresford's thesis provides clear and detailed documentation of the search for new and unpredicted states of matter. It is focused on searches for new phenomena in jet final states; jets of sub-atomic particles appear in the ATLAS detector due to a linear increase in the strength of the strong force as quarks or gluons separate from each other, causing an increase in the energy density. When that energy density is high enough, particles with a colour charge are created from free space and travel in roughly the same direction as the original progenitor quark or gluon, and thus a "jet" is formed.

Just prior to Lydia's research, the LHC significantly upgraded its available energy in the centre of mass, from 8 to 13 TeV. Such an increase means that with relatively small amounts of data, large gains in the search region for new states of matter become immediately possible. This motivated performing a search for heavy new particles which decay into a pair of jets, a "dijet". The basic idea behind the high-mass search is deceptively simple. By adding up the energy and momentum produced by the most energetic pair of jets observed in the detector, one can find the "mass" of the particle that produced those jets. Most of the time the assumption that there is indeed a single particle which decayed to produce the jets is incorrect; in which case a continuous, falling spectrum as a function of this "dijet mass" results. However, should the assumption be correct, then data will build at a particular dijet mass, producing an excess which increases with time as the machine runs. Such a state could exist in a wide range of possible theoretical extensions to the Standard Model. One such extension could be the existence of a dark matter mediator particle which, if found, would give us the chance to create and study dark matter.

In addition to the search for high-mass states, a new technique has been utilized and presented in this thesis to re-explore the lower dijet mass regions. Some of the most stringent constraints at low dijet mass had previously been set by existing experiments (e.g., CDF and D0 at Fermilab, and UA1/UA2 at CERN). By triggering on a single high energy jet or on a single high energy photon and then performing a dijet analysis on cases where at least two additional jets are present, dijet masses down to 200 GeV could be probed. This allowed stronger constraints to be set in this region than was possible previously.

Even if there are no indications of an excess that would be caused by a new particle, as is the case for the searches presented in this thesis, its absence helps us to limit the kind and number of theories that aspire to extend or supplant the Standard Model. Our understanding of the Universe, as a result of these searches, is therefore enriched and enhanced nonetheless.

This body of work shows in great detail how such searches are carried out and presents the challenges that one encounters within such research. This thesis is an important read for those who wish to learn more about the experimental aspects of searches and limit setting. The thesis explains how jets are constructed in the ATLAS experiment, how their energy is determined, and the uncertainties on that energy estimated. The process of determining where the most significant deviation from a smooth background occurs, and whether this deviation is significant is explained. This thesis goes further by comparing the limit on massive particle production obtained to several theoretical models. In summary, this thesis is an essential read for anyone in particle physics who wishes to gain insights into how searches for new physics are performed in the modern post-Standard Model era.

Oxford, UK
July 2018

Prof. B. Todd Huffman

Abstract

This thesis presents three searches for new resonances in dijet invariant mass spectra. The spectra are produced using $\sqrt{s} = 13$ TeV proton–proton collision data recorded by the ATLAS detector. New dijet resonances are searched for in the mass range from 200 GeV to 6.9 TeV in mass. Heavy new resonances, with masses above 1.1 TeV, are targeted by a high-mass dijet search. Light new resonances, with masses down to 200 GeV, are searched for in dijet events with an associated high momentum object (a photon or a jet) arising from initial state radiation. The associated object is used to efficiently trigger the recording of low-mass dijet events. All of the analyses presented in this thesis search for an excess of events, localised in mass, above a data-derived estimate of the smoothly falling QCD background. In each search, no evidence for new resonances is observed, and the data are used to set 95% C.L. limits on the production cross section times acceptance times branching ratio for model-independent Gaussian resonance shapes, as well as benchmark signals. One particular benchmark signal which is considered in all of the searches is an axial-vector Z' dark matter mediator model whose parameter space is reduced due to the results presented in this thesis.

Acknowledgements

There are many people who have contributed in some way to the completion of this thesis, and I am sincerely grateful to all of you. I would like to name just a few of them here.

First and foremost, I would like to thank my supervisors, Todd Huffman and Çiğdem İşsever, for your endless support and encouragement, the helpful discussions that we have had and the advice that you have given me. It has been a pleasure to work with you both.

I would like to thank Elizabeth Gallas for her help and support during my qualification task and all members of the Oxford ATLAS group for providing me with helpful feedback and the perfect working environment.

I have had the privilege of working in several fantastic analysis groups during my DPhil; I would like to thank all members of the dijet analysis team, the dijet + ISR analysers and the TLA team—it has been a pleasure to work with you all. In particular, I would like to thank Katherine Pachal, for imparting some of her extensive statistical knowledge on me, for helping me to get to grips with the statistics code, and for answering my questions at all times of the day and night; and Antonio, Caterina, Francesco, Gabriel, James, Jeff, John, Karol, Laurie, Lene, Prim, for all of the helpful discussions we have had, and for answering my many questions. Thanks also to Frederik Beaujean for your helpful explanations of the BAT package.

I am extremely grateful to Shaun Gupta, Caterina Doglioni and Jonathan Bossio for teaching me all about jet punch-through, and to Steven Schramm, Dag Gilbert and Andy Pilkington for all of your helpful advice about the jet punch-through uncertainty.

I must also express my gratitude to STFC for funding my DPhil and making this thesis possible, also to Wolfson College for funding my conference trips and providing me with the perfect living environment.

My time in Oxford and at CERN would not have been the same without all of the wonderful people I have been able to share it with. I would like to thank my friends at the Oxford physics department, in particular, Anita, FengTing, Fikri,

Jack, James, Jon, Kathryn, Mariyan, Mark, Pete, Stephen, Tim, Will F. and Will K. for all of the discussions we have had, both physics and non-physics related, and for all of the lunches, tea breaks and laughs shared. I would also like to thank the whole of the LTA for making my time at CERN great, and all my friends from Wolfson College, in particular, Claudia, Corina, Luke, Maurits and Nina; I have many fond memories of all the BOPs attended, nights at Wolfson bar, tea breaks after dinner and everything else.

Claudio, I can't thank you enough; you have been there for me through it all, providing infinite support, encouragement and advice. I am looking forward to everything that life has in store for us and all the adventures yet to come.

To Mum, Dad and Jordan, thank you for being there for me at all times, not only during my DPhil, but throughout my whole life. Your support and encouragement has helped me to get to where I am today, and I am immensely grateful for this. This is why this thesis is dedicated to you.

Contents

Chapter 1
Introduction

In the quest to further our knowledge about what the universe is made from, and the forces which hold it together, humans have built particle accelerators with increasing centre-of-mass energies ($\sqrt{s}$), to try to discover new particles and interactions. In 2015, the Large Hadron Collider (LHC) based at the European Organisation for Nuclear Research (CERN), managed to reach the highest centre-of-mass energy ever achieved by a particle accelerator, $\sqrt{s} = 13$ TeV. This allows us to explore higher energy scales, and to search for new heavy particles whose production may have previously been kinematically forbidden.

The proton-proton (pp) collisions delivered by the LHC are recorded by detectors, producing vast amounts of experimental data. Two such detectors are the ATLAS and CMS general purpose detectors. In 2012, the ATLAS and CMS experiments completed the search for Standard Model (SM) particles with the discovery of the Higgs boson [1, 2]. The focus of these two experiments is now to search for new Beyond Standard Model (BSM) particles and interactions, as the Standard Model is an incomplete theory. This thesis describes three analyses which use data recorded by the ATLAS detector to search for BSM phenomena in the two jet (dijet) final state.

As the LHC is a hadron collider, there are several reasons why the dijet final state is an interesting and important final state to search in. Firstly, new particles directly produced in the collisions must couple to quarks and gluons, therefore they are also expected to be able to decay to quarks and gluons, producing jets in the final state. For many BSM models, the dijet production rate is large, even at relatively large fractions of the centre-of-mass energy. Therefore, by using the dijet final state, high mass scales can be probed with relatively little data.

This thesis focuses on searches for resonances in the dijet final state from the production of new particles. As the invariant mass spectrum of dijet events is predicted by Quantum Chromodynamics (QCD) to be smoothly falling with increasing dijet mass, resonances can be searched for by looking for an excess of events, localised in mass, above the smoothly falling background. Three analyses are described in this

L. A. Beresford, *Searches for Dijet Resonances*, Springer Theses,
https://doi.org/10.1007/978-3-319-97520-7_1

thesis, one which targets the high mass region of the dijet invariant mass spectrum, and two which target the low mass region:

- High mass dijet analysis (1.1 TeV and above)
- Dijet + γ analysis (200–1500 GeV)
- Dijet + jet analysis (300–600 GeV)

The dijet $+\gamma$ and dijet + jet analyses are collectively referred to as the dijet + Initial State Radiation analyses (or dijet + ISR analyses).

This thesis is organised as follows: In Chap. 2, the Standard Model is introduced, with particular focus on proton-proton collisions, Quantum Chromodynamics, and the formation of jets. The limitations of the Standard Model and possible extensions are also introduced, with a focus on dark matter, and the motivation for performing the analyses documented in this thesis is outlined. The LHC and the ATLAS detector are described in Chap. 3. Chapter 4 details how jets are reconstructed and calibrated, together with how the systematic uncertainties for these calibrations are derived. Particular focus is given to the description of the uncertainty due the jet punch-through correction. The reconstruction of photons is also described. The production of the dijet invariant mass spectra for the three analyses described in this thesis is outlined in Chap. 5. The search techniques, validations, and results for each analysis are given in Chap. 6. In the absence of observing new phenomena, limits are set on benchmark models, and on model-independent Gaussian shapes, the techniques and results for the limit setting are presented in Chap. 7.

References

1. ATLAS Collaboration (2012) Observation of a new particle in the search for the standard model Higgs boson with the ATLAS detector at the LHC. Phys Lett B 716.1:129. https://doi.org/10.1016/j.physletb.2012.08.020, arXiv:1207.7214 [hep-ex]
2. CMS Collaboration (2012) Observation of a new boson at a mass of 125 GeV with the CMS experiment at the LHC. Phys Lett B 716.1:3061. https://doi.org/10.1016/j.physletb.2012.08.021, arXiv:1207.7235 [hep-ex]

Chapter 2
Theoretical Background

The theoretical description of particles and their interactions is provided by the Standard Model of particle physics, also referred to as the Standard Model in this thesis. This theory was finalised in the 1970s, and has been extremely successful, with experimental results confirming the predictions of the Standard Model to increasing degrees of precision [1]. Despite the huge successes of the Standard Model, there are many reasons to believe that this theory is incomplete, and that there is new physics 'beyond' the Standard Model.

In this chapter, a brief introduction to the Standard Model will be provided in Sect. 2.1, with particular focus on the particle content and the fundamental forces it describes. Section 2.2 focuses on features of the strong interaction, and the resulting phenomena of jet formation. Section 2.3 describes the motives for searching beyond the Standard Model, and provides examples of new physics models which could decay into pairs of jets. Finally, in Sect. 2.4, the analyses described in this thesis are introduced, together with the reasons for performing them.

2.1 The Standard Model of Particle Physics

The Standard Model (SM) provides a theoretical description of all the known elementary particles, and their interactions via three of the four fundamental forces: the electromagnetic (EM) interaction, the weak interaction, and the strong interaction (the gravitational force is not included in the Standard Model). It is a quantum field theory, with particles corresponding to excitations of fields. The dynamics of these fields are encoded in the Standard Model Lagrangian. A brief overview of the Standard Model will be provided here, for a detailed description see [2, 3].

The elementary particles described by the Standard Model can be divided into two classes: particles with half integer spin (intrinsic angular momentum) are referred to as *fermions*, and particles with integer spin are referred to as *bosons*. A summary of all the known elementary particles is displayed in Fig. 2.1.

L. A. Beresford, *Searches for Dijet Resonances*, Springer Theses,
https://doi.org/10.1007/978-3-319-97520-7_2

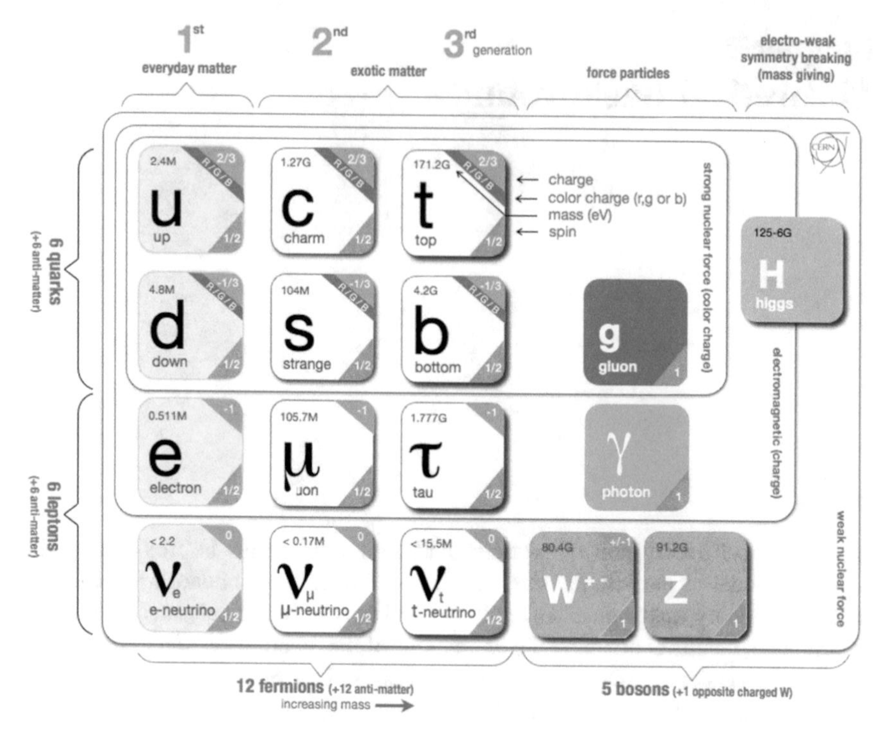

Fig. 2.1 This figure [4] shows the elementary particles contained in the Standard Model, together with a summary of their basic properties

The fermions in the Standard Model make up the visible matter in the universe. They are all spin $\frac{1}{2}$ particles, and can be classified further into quarks and leptons. One key distinction between quarks and leptons is that quarks can interact via the strong force, and leptons cannot. There are six flavours of quark, and they are arranged into pairs, referred to as *generations*. Quarks possess a fractional electric charge of $+\frac{2}{3}$ for *up type* quarks, and $-\frac{1}{3}$ for *down type* quarks. Similarly, there are six flavours of leptons, also arranged into generations. The charged leptons possess a charge of -1, and the neutral leptons are called neutrinos. For each of the fermions there exists a corresponding anti-fermion, with the same mass, but opposite quantum numbers.

The Standard Model is based on the gauge group SU(3) $\times$ SU(2) $\times$ U(1), where SU(3) corresponds to the strong force, SU(2) corresponds to the weak force, and U(1) corresponds to the electromagnetic force. The requirement for the Standard Model Lagrangian to be unchanged under local gauge transformations introduces gauge bosons associated with each gauge group. The gauge bosons in the Standard Model are all spin 1 particles and are the mediators of interactions. The mediator of the electromagnetic force is the photon. The photon is massless, electrically neutral, and interacts with charged particles. The mediators of the weak force are the W^+, W^- and Z^0 bosons, with the superscript indicating the charge of the boson. These massive bosons can interact with fermions and also have self-interactions. Finally, the mediator of the strong force is the gluon. The gluon is massless, electrically neutral and can interact with quarks, and with itself.

The final particle in the Standard Model is the Higgs boson. The Higgs boson is a massive, scalar boson, and is produced as a result of the Higgs mechanism [5, 6], which breaks the SU(2) $\times$ U(1) symmetry, generating the masses of fermions and bosons. The Higgs boson interacts with all massive particles, and therefore is predicted to have self-interactions. This completes the particle content of the Standard Model.

2.2 Quantum Chromodynamics

Quantum Chromodynamics (QCD) is the theory of the strong force (for a detailed review of QCD see [7]). The mediator of the strong force, the gluon, can interact with any particles possessing *colour charge*. Colour charge is the conserved quantum number of strong interactions and is analogous to the electric charge in EM interactions. Quarks possess a colour charge of red, green, or blue, and anti-quarks possess a colour charge of anti-red, anti-green, or anti-blue, hence, quarks and anti-quarks can interact with gluons. There are in fact eight types of gluons, each possessing a different superposition of colour and anti-colour charge, and hence, gluons can interact with themselves.

Examples of QCD interaction vertices are shown in Fig. 2.2. The interaction vertex between a quark, an anti-quark, and a gluon denoted $q, \bar{q}$, and g, respectively, is shown in Fig. 2.2a. Figure 2.2b shows the three gluon self-interaction vertex. Figure 2.2b shows the four gluon self-interaction vertex. The strength of the interaction at the vertices in Fig. 2.2a, b is proportional to the square root of the strong coupling constant $\sqrt{\alpha_s}$ (or equivalently to $g_s = \sqrt{4\pi\alpha_s}$). In Fig. 2.2b the strength of the interaction is proportional to α_s (or equivalently to g_s^2).

2.2.1 Confinement and Asymptotic Freedom

A unique feature of QCD is the relationship between the strong coupling constant α_s and the momentum transfer of the interaction Q. Figure 2.3 shows α_s as a function of Q, illustrating the fact that in reality α_s is not a constant, but *runs* with Q. In contrast to the electromagnetic coupling constant, which increases as a function of Q, the strong coupling constant decreases as a function of Q, with $\alpha_s \to 0$ as $Q \to \infty$. This phenomenon is referred to as *asymptotic freedom*. The implication of asymptotic freedom is that when a probe with sufficiently high energy interacts with a composite particle, the constituent quarks and gluons behave like free particles and can be resolved [9]. Conversely, for low momentum transfer the quarks are tightly bound by the gluons, forming colour neutral, composite particles; this phenomenon

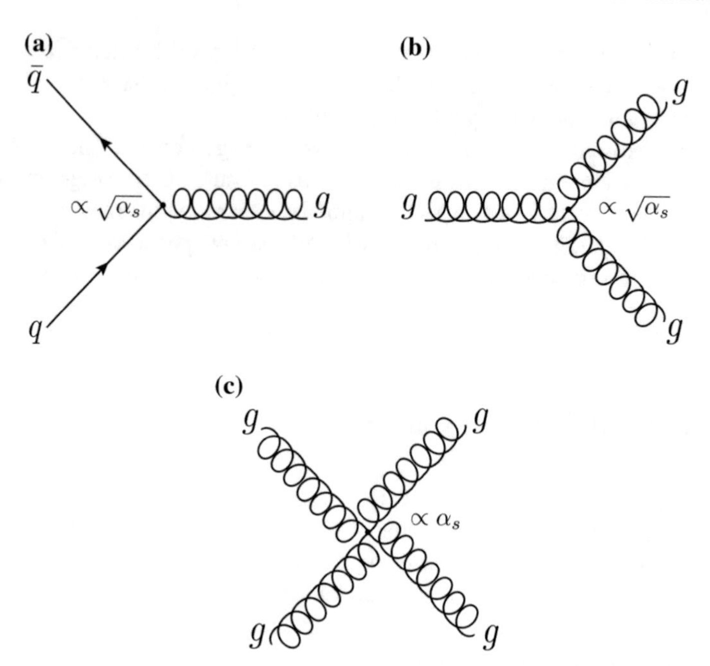

Fig. 2.2 Figure **a** shows the annihilation of a quark, anti-quark pair $q\bar{q}$ producing a gluon g. Figure **b** shows the splitting of a gluon into a pair of gluons. Figure **c** shows the self-interaction between four gluons. The strength of the interaction at the vertices in Figure **a** and **b** is proportional to $\sqrt{\alpha_s}$. In Figure **c**, the strength of the interaction at the vertex is proportional to α_s

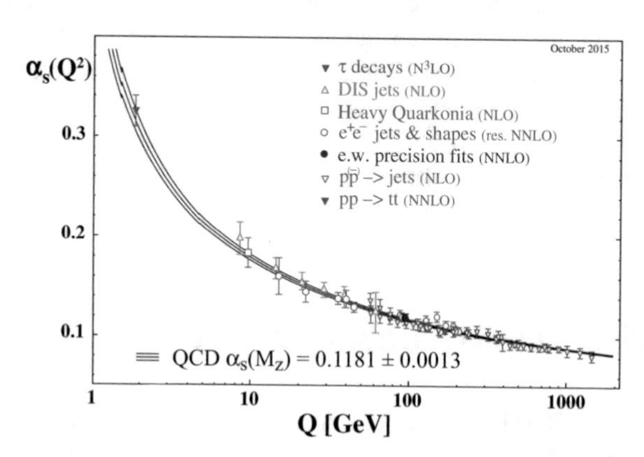

Fig. 2.3 The strong coupling constant α_s is shown as a function of the momentum transfer of the interaction Q. Each of the data points is from an experimental measurement of α_s, and the world average is shown by the grey lines. This figure is taken from [8]

is referred to as *confinement*. The composite particles formed from quarks are called *hadrons*. Hadrons composed of a quark and an anti-quark pair $q\bar{q}$ are called *mesons*, and hadrons composed of three quarks qqq are called *baryons* ($\bar{q}\bar{q}\bar{q}$ is called an *anti-baryon*).

The magnitude of the strong coupling constant also has important consequences for the calculation of physical quantities. In the high Q region where α_s is small ($\ll 1$) perturbation theory can safely be applied and observables can be expanded in a power series with increasing orders of α_s [10]. However, in the low Q region ($Q < 1\,\text{GeV}$) α_s is large ($\mathcal{O}(1)$) and perturbation theory can no longer be utilised; the region is said to be non-perturbative. The impact of this will be discussed in Sect. 2.2.3.

2.2.2 Jet Formation in Proton–Proton Collisions

In high energy collisions involving hadrons the processes of asymptotic freedom and confinement can lead to the formation of collimated showers of colour neutral particles, referred to as jets. The formation of jets in proton–proton collisions will now be described, following [11, 12]. An illustration of jet formation is shown in Fig. 2.4.

The two incoming protons can be thought of as bags of *partons* (quarks and gluons). When these high energy protons collide, a high Q interaction occurs between two partons, referred to as the *hard scattering process*. In the hard scattering process a short lived *resonance* particle could be created, for example, a Z boson. Alternatively, a standard QCD process could occur, for example, gluon-gluon scattering $gg \rightarrow gg$ as shown in Fig. 2.4. The hard scattering process can result in the production of partons, either through the decay of a created resonance, or through standard QCD processes.

In addition to the hard scattering process, the incoming partons can radiate particles, for example, photons or gluons, producing *Initial State Radiation* (ISR). Similarly the outgoing partons can produce *Final State Radiation* (FSR). The interactions of partons in the protons which were not involved in the hard scattering process form the so-called *Underlying Event* (UE).

At sufficiently high energies, the partons produced in all of these interactions can split to produce more partons (for example, $g \rightarrow gg$, $g \rightarrow q\bar{q}$) forming a parton shower. The partons produced in the splitting process are typically produced at a small angle to the original parton; hence, the showers are collimated in the direction of the original parton. Once the produced partons move further apart and reach lower energies they are no longer asymptotically free and confinement takes over. The partons bind together to form hadrons in a process called *hadronisation*. Hadronisation is a non-perturbative process and cannot be explained from first principles. Instead models are used to describe this process. More details about such models will be provided in the next section.

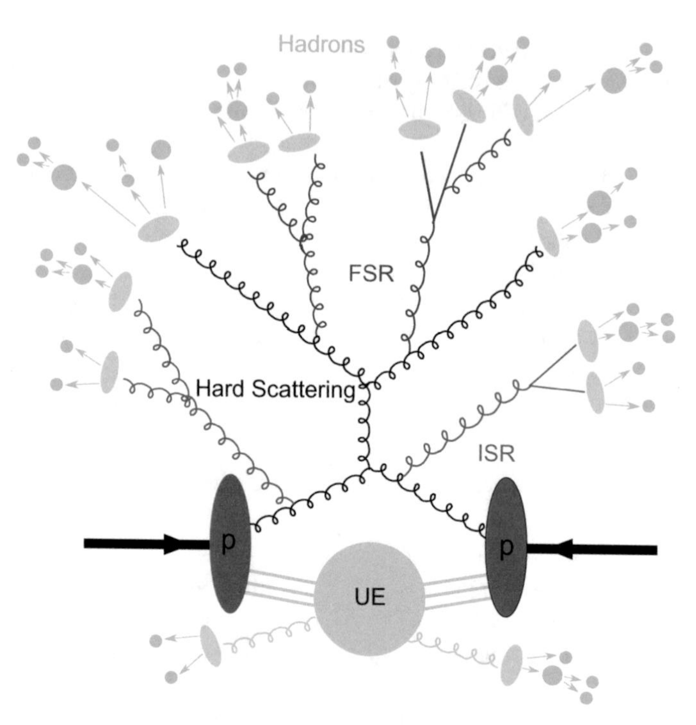

Fig. 2.4 This figure illustrates the stages of jet formation from the collision of two protons p. The hard scattering process of $gg \rightarrow gg$ is illustrated in black in the centre. Initial State Radiation (ISR) from the incoming partons involved in the collision is shown in pink, and Final State Radiation (FSR) from the outgoing partons is shown in blue. The partons in the proton which are not involved in the hard scattering process can interact, producing the Underlying Event (UE) shown in orange. The transition from partons to hadrons is represented by the light grey ovals, and the resulting hadrons are shown in green. This figure is adapted from [13, 14]

The end result is that each of the partons forms a collimated shower of hadrons in approximately the same direction as the original parton which initiated the shower; this group of hadrons is called a jet. Details about the algorithms used to reconstruct jets will be given in Chap. 4.

2.2.3 *Event Simulation*

From an experimental perspective, one must be able to use the Standard Model or Beyond Standard Model theory to make accurate predictions of the outcome of experiments. Often such predictions are in the form of *Monte Carlo simulations*, in which all the stages listed in the previous section are simulated on an event-by-event basis, producing a set of particles for each event. For a large sample of generated events, the underlying distributions reflect the probability distributions of the theory being simulated [15].

Monte Carlo simulations are utilised in many aspects of experimental analysis, including calibration, analysis optimisation, and determination of systematic uncertainties. Many analyses also utilise Monte Carlo simulations to model the expected *background* from Standard Model processes. However, for the analyses described in this thesis, the background estimate is determined using an empirical fit to the data, as will be described in Chap. 6. Background Monte Carlo samples are used to test the background estimation procedure though and Monte Carlo simulations are utilised to model the predicted *signal* from Beyond Standard Model processes.

Monte Carlo simulations are a broad and complicated topic, and only a very brief summary of the key steps will be provided here. For a more detailed explanation see [12, 14], on which this section is based.

Hard Scattering

At the heart of a Monte Carlo simulation is a calculation of the cross-section for the hard scattering process being simulated, reflecting the probability for the process to occur. The calculation of the cross-section relies on the QCD factorisation theorem [16], which allows us to separate (factorise) the perturbative hard scattering partonic cross-section, from the non-perturbative low momentum interactions, which are described by *Parton Distribution Functions* (PDFs).

The partonic cross-section is proportional to the square of the matrix element (transition amplitude) for the process being simulated, and can be expanded using perturbation theory with the coupling constant for the interaction as the expansion parameter. A finite number of terms is used to approximate the partonic cross-section. Each of the terms in this expansion can be calculated using Feynman diagrams with a corresponding set of Feynman Rules derived from the Lagrangian of the Standard Model or Beyond Standard Model theory being calculated. For details about calculations using Feynman diagrams see [17].

Figure 2.5 shows examples of Feynman diagrams for partonic processes which result in the production of a pair of jets (dijet). Figure 2.5a shows an example of a *t-channel* process, and Fig. 2.5b shows an example of an *s-channel* process. QCD background is dominated by t-channel processes, which have an enhanced cross-section for small angle scattering in the centre of mass frame, as described in [11]. The diagrams shown correspond to leading order in perturbation theory as they represent the lowest order in α_s needed to produce a $2 \rightarrow 2$ process. If an additional quark or gluon is emitted, or a virtual loop is included, then this is referred to as next-to-leading order [18].

Matrix element generators are used to generate and compute the relevant Feynman diagrams. The results are then convoluted with the PDFs and Monte Carlo methods are used to compute the integrals in the cross-section calculation.

Parton Distribution Functions

The PDFs give the probability for a parton to carry a fraction x of the longitudinal momentum of the incoming proton, and have a dependence on the momentum

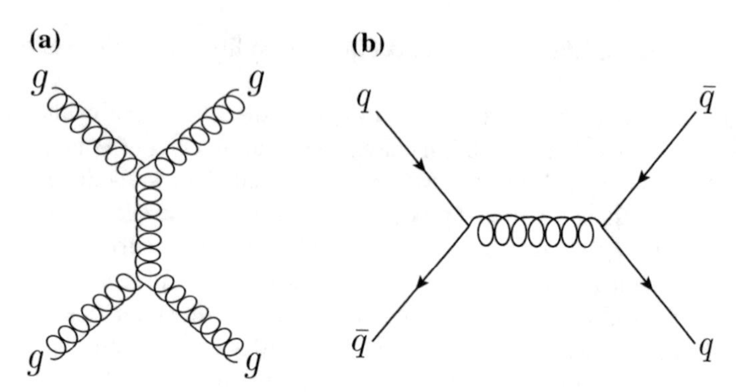

Fig. 2.5 These figures show examples of Feynman diagrams for the production of a pair of partons via **a** t-channel gluon-gluon scattering $gg \to gg$; and **b** s-channel quark, anti-quark annihilation and production $q\bar{q} \to q\bar{q}$

transfer of the hard scattering process Q. The PDFs allow us to translate between the momentum of the protons and the momentum of the partons taking part in the hard scattering process. Since PDFs involve the non-perturbative region of α_s, they are derived empirically. Several PDF sets are available from different collaborations.

Parton Shower

As previously described, the hard scattering process does not occur in isolation, and both the incoming and outgoing partons can radiate additional particles. Modelling the full $2 \to n$ process using matrix elements would be extremely complex and time consuming. Therefore, an alternative approach is taken. The additional radiation is simulated using a *parton shower generator*, and the hard scattering process is then merged with the parton shower. The parton shower approach approximates shower formation using *splitting functions*, which give the probability for partons to 'split' and produce new particles. This splitting continues until a cut-off scale in Q is reached, typically 1 GeV.

Underlying Event

In addition to the hard scattering process and the ISR and FSR, the underlying event can also produce partons. The underlying event refers to the interactions of the partons in the two colliding protons which were not involved in the hard scattering process. Modelling the underlying event is complex and will not be described here, for details about the modelling of the underlying event see [19]. The simulation of the underlying event is typically performed by the parton shower generator. The parton shower generator has 'tunable parameters' which can be set to improve the modelling of the parton shower and underlying event; optimised sets of these parameters are referred to as Monte Carlo *tunes*.

Hadronisation

As the partons produced in the previous steps move further apart and lose energy, they begin to hadronise. As previously mentioned, due to the non-perturbative nature of hadronisation, this process cannot be explained from first principles, and models are employed to simulate hadronisation. There are two main models of hadronisation in use today. These are the Lund string model [20] and the cluster model [21].

In the Lund string model, quarks are joined together by strings which stretch as the quarks separate. These strings can snap, producing a $q\bar{q}$ pair at the end of the two new strings. This process continues until sufficient energy is no longer available, and the strings have fragmented into hadrons.

In the cluster model, gluons split into $q\bar{q}$ pairs and form colour neutral clusters with neighbouring partons. These clusters are then decayed into hadrons.

Hadronisation is typically performed by the parton shower generator. Two popular generators are the PYTHIA generator [22], which utilises the Lund string model, and the SHERPA generator [23], which utilises the cluster model. PYTHIA and SHERPA are both general purpose event generators and can also be used to generate matrix elements.

The final step is to simulate the decay of any unstable hadrons which have been produced.

Although each step in the Monte Carlo production has been described separately, the processes are all closely interlinked. The Monte Carlo generated is said to be at *truth level*, meaning that it is what we would obtain with a perfect detector. To obtain *reconstructed level* Monte Carlo, i.e. the output obtained when the simulated particles interact with the ATLAS detector, the truth level particles and additional particles from pile-up interactions (interactions in other proton–proton collisions, described in full in the next chapter) are put through a detector simulation using GEANT4 [24] within the ATLAS simulation infrastructure [25]. The energy deposits left in the simulated detector can then be reconstructed to create particles in the same manner as the data reconstruction, which will be described in Chap. 4.

2.3 Beyond the Standard Model

The Standard Model is able to provide accurate predictions for numerous interactions involving three of the fundamental forces, providing impressive agreement with experimental results. However, the Standard Model does not explain several experimentally observed phenomena, indicating that it is not a complete theory.

As previously mentioned, the gravitational force is not included in the Standard Model. At energies well below the Planck scale ($\sim 10^{16}$ TeV), the gravitational force is many orders of magnitude weaker than the other three fundamental forces, and interactions via the gravitational force can be neglected [26]. However, as we approach the Planck scale, when gravitational interactions become comparable in strength to the other forces and can no longer be neglected, the Standard Model breaks down.

A further deficiency of the Standard Model is that it does not incorporate neutrino masses. In the Standard Model, neutrinos are considered to be massless, however, the observation of neutrinos oscillating from one flavour to another indicates that they have a non-zero mass.

Another key shortcoming of the Standard Model is that it does not provide a suitable dark matter candidate, despite compelling experimental evidence for the existence of dark matter. A brief summary of this evidence will be given, for further details see [8, 27, 28], on which this section is based. In the 1930's, a surprising observation was made by F. Zwicky [29]. Based on the amount of luminous matter, Zwicky estimated that the velocity dispersion of the galaxies in the Coma cluster should be $\sim$80 km s^{-1}. However, his measurements indicated that the velocity dispersion was in fact $\sim$1000 km s^{-1}. With the profound implication that there is 'non-luminous' (dark) matter providing the additional gravitational attraction needed to hold the galaxies within the cluster. Since then, numerous measurements have provided powerful support for dark matter.

In 1980, Rubin, Ford and Thonnard measured rotation curves for 21 spiral galaxies [30]. The rotation curves show the orbital velocity of stars and gas clouds in each galaxy as a function of their radial distance from the centre. From Newtonian dynamics, the following relationship between velocity $v(r)$ and radial distance r is expected:

$$v(r) = \sqrt{\frac{GM(r)}{r}}, \tag{2.1}$$

where G is the gravitational constant and $M(r)$ is the mass within radius r. Beyond the galactic disk, $M(r)$ is expected to be constant, and velocity would fall off as $v(r) \propto \frac{1}{\sqrt{r}}$. However, the observed rotation curves displayed a slow rise in velocity beyond the galactic disk, as sketched in Fig. 2.6. These results indicate that the galactic disk is surrounded by a large halo of dark matter, providing additional mass.

One could imagine that these observed effects could be due to modified gravitational forces, affecting the rotations of stars and galaxies, rather than particulate

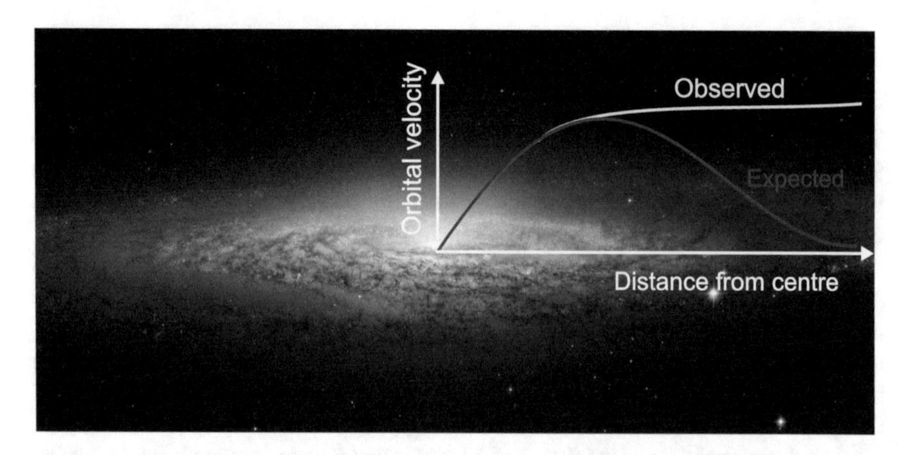

Fig. 2.6 This figure sketches the expected galactic rotation curve in red, assuming only luminous matter, and the observed rotation curve in yellow, which does not drop off beyond the galactic disk. Figure adapted from [31, 32]

Fig. 2.7 The bullet cluster shows the collision between two clusters of galaxies. The hot gas, shown in pink, is measured via its emission of X-rays, and the inferred dark matter, shown in purple, is measured via gravitational lensing [33]

dark matter. However, the famous ‘bullet cluster’, shown in Fig. 2.7 is difficult to explain using theories of modified gravity. The figure shows the collision between two clusters of galaxies. When these clusters collide, the hot gas from each cluster (shown in pink) interacts and slows down, whereas the inferred dark matter (shown in purple), passes through the gas clouds with minimal interaction [33]. The hot gas is observed via its X-ray emissions, which are recorded by NASA’s Chandra X-ray Observatory, and the non-luminous matter is measured via gravitational lensing [33].

The evidence presented so far strongly supports the existence of dark matter, but does not quantify the relative abundance of dark matter. The latest Planck space mission has provided the most precise measurement of anisotropies in the Cosmic Microwave Background (CMB) to date [34]. The CMB is relic radiation from the early universe. Measurements of the CMB anisotropies (temperature fluctuations) can be compared to the predictions from cosmological models, constraining the parameters of the model. The standard cosmological model was considered, called the ΛCDM model, where Λ is the cosmological constant representing dark energy, and CDM stands for Cold Dark Matter, i.e. dark matter particles which move slowly with respect to the speed of light. One of the parameters in this model is the relative abundance of dark matter, which is determined to be 26%. The remaining mass-energy in the universe is divided between visible matter (5%) and dark energy (69%).

These astrophysical observations illustrate that there is compelling evidence for the existence of dark matter. One of the fundamental aims of particle physics is to identify and characterise dark matter.

In order to address some of the outlined deficiencies of the Standard Model, and to address more ‘aesthetic’ issues, such as why the quarks and leptons are arranged into

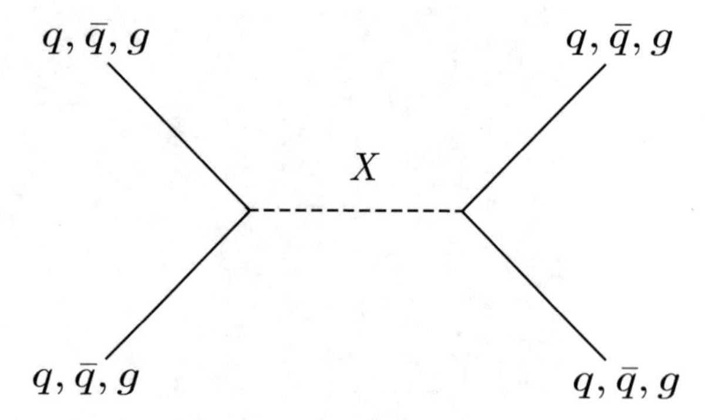

Fig. 2.8 Feynman diagram showing the s-channel production of a new resonance X, and its subsequent decay into a pair of partons

three generations with a mass hierarchy, theorists have developed many new Beyond Standard Model (BSM) theories. The analyses presented in this thesis aim to search for BSM physics via the s-channel production of a new resonance, as illustrated in Fig. 2.8, resulting in a localised excess in the dijet invariant mass spectrum at the mass of the resonance. A model independent approach is taken, in which we search for new resonances from any BSM model. In the absence of observing new BSM physics, several models are utilised as 'benchmarks' to assess the progress of the analysis, and to show the phase space excluded for these models by our analyses. Additionally, some of these benchmark models are used in the optimisation of the analysis. A brief summary of these benchmark BSM theories will now be presented, together with the motivations behind them. Note that the Z' dark matter mediator model is utilised in all of the analyses described in this thesis, and the other models described here are only utilised in the high mass dijet analysis.

2.3.1 Z' Dark Matter Mediator

There is substantial evidence that a large fraction of our universe is composed of dark matter. This evidence suggests that dark matter is particulate, neutral, and interacts via the gravitational force [8]. A well motivated cold dark matter candidate is a *Weakly Interacting Massive Particle* (WIMP). It can be shown that a $\mathcal{O}$(GeV–TeV) particle with weak scale couplings could produce the observed dark matter abundance. This phenomenon is known as the *WIMP miracle* [35].

In the analyses described in this thesis, we search for a weak scale dark matter mediator particle Z', which links the Standard Model particles to dark matter particles. By searching for the dark matter mediator through its decay to quarks, we can access phase space which is kinematically inaccessible for other analyses which search for dark matter particles directly, making our approach complementary [36]. The model utilised is recommended by the ATLAS and CMS Dark Matter Forum to maintain consistency between searches, and is described in detail in [37]. The new

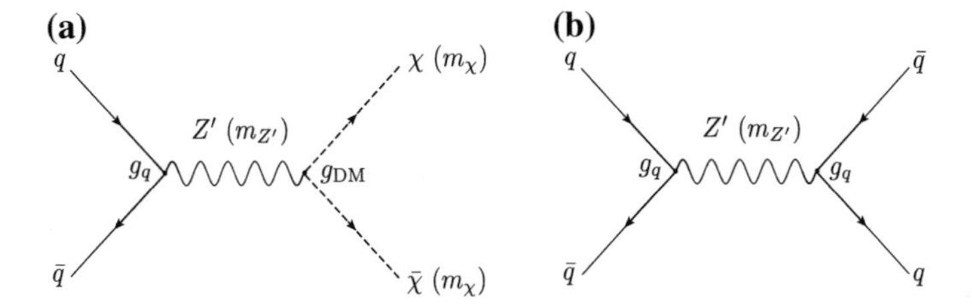

Fig. 2.9 Feynman diagrams showing the production of a Z' with mass $m_{Z'}$, and its subsequent decay to **a** a pair of dark matter particles $\chi\bar{\chi}$, each with mass m_χ; and **b** a pair of quarks $q\bar{q}$. The Z' coupling to quarks is denoted g_q, and the Z' coupling to dark matter particles is denoted $g_{\rm DM}$. Figure adapted from [37]

particles and parameters introduced by the model are illustrated in Fig. 2.9. Figure 2.9a shows the decay of the Z' particle to a pair of dark matter particles $\chi\bar{\chi}$, and Fig. 2.9b shows the decay of the Z' particle to a $q\bar{q}$ pair, which would result in the production of a dijet resonance.

The dark matter particle χ is a fermion with mass m_χ, and the dark matter mediator particle Z' is a spin 1 boson (from the addition of a U(1) gauge group) with mass $m_{Z'}$ and axial-vector couplings to quarks and to χ. The Z' coupling to quarks is universal for all quark flavours and is denoted by g_q, and the coupling to the dark matter particles is denoted by $g_{\rm DM}$. The coupling of the Z' to leptons is forbidden, making it *lepto-phobic*.

Note that previous dijet analyses utilised a baryonic Z' model. References and comparisons to this model will be made in this thesis, so a very brief description of the key features is given (for full details see [38]). The baryonic Z' is also lepto-phobic, but has no coupling to dark matter and has vector couplings to quarks. For this model, the coupling to quarks is denoted by g_B, and g_B is related to g_q for the Z' model utilised in this thesis by a factor of 6 ($g_q = \frac{g_B}{6}$), due to a difference in the definition of the coupling (the baryonic Z' model includes the number of quarks in the definition of the coupling).

2.3.2 *Heavy W′ Boson*

Heavy spin 1 bosons are predicted in many theories with additional gauge groups. The particular case we consider here is the charged W' bosons from the *Sequential Standard Model* [39], in which the new bosons are heavier versions of the Standard Model W bosons, with the same couplings.

2.3.3 Excited Quarks

Excited quarks q^* [40, 41] are a typical signature of composite quark models. In such models quarks are not point-like, but are in fact a bound state of constituent particles. If true, this could help to explain the generation structure and mass hierarchy of quarks. Excited quark models have been used in many previous dijet resonance searches. Hence, the inclusion of this model facilitates the comparison of results.

2.3.4 Quantum Black Holes

The fundamental scale of gravity, the Planck scale, is 16 orders of magnitude above the Electroweak scale; a phenomenon referred to as the *hierarchy problem*. As a consequence, the gravitational force is much weaker than the other fundamental forces. A possible explanation for this could be that the gravitational force is of similar strength to the other forces, but it appears to be weak as it can propagate in additional space-time dimensions. Two popular models of extra dimensions are the ADD (Arkani-Hamed Dimopoulos Dvali) model [42, 43], which introduces 6 large extra dimensions, and the RS (Randall Sundrum) model [44], which introduces 1 additional warped dimension. Such extra dimensions can be searched for at the LHC through their effects on gravity. The extra dimensions can lower the fundamental scale of gravity M_D to the TeV scale, with the consequence that micro black holes could be formed in parton-parton collisions at the LHC [45]. Micro black holes with masses $\sim M_D$ are called quantum black holes and could decay to 2-body final states, as described in [46]. This gives rise to a resonance shape in the dijet invariant mass distribution, due to the combination of a production turn on effect once the mass threshold for black hole production M_{Th} is passed, coupled with strongly falling parton distribution functions, resulting in an excess of events at $\sim M_{Th}$.

2.4 Dijet Resonance Searches and Motivation

In order to guide and test the theoretical extensions to the Standard Model, experimentalists conduct searches for new BSM particles and interactions, and set exclusion limits on physical attributes (e.g. mass or production cross-section) of benchmark models. There are many different final states and experimental signatures that one could use to conduct searches for BSM physics. This thesis focuses on the search for resonances in dijet final states. As mentioned in the introduction, there are many reasons why the dijet final state is interesting. A more detailed explanation of the motivation for using the dijet final state will be given here, with references to the limits achieved by previous analyses, and explanations about why searching for

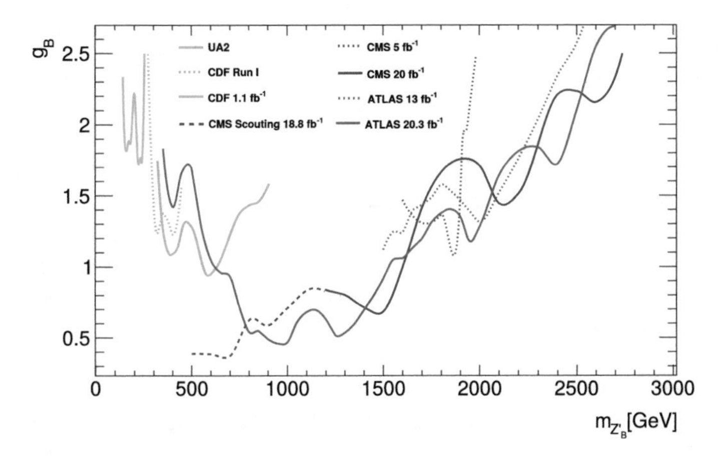

Fig. 2.10 This figure shows the limit contours for a variety of different experiments (UA2, CDF, CMS and ATLAS), with different dataset sizes and centre-of-mass energies. The limits are in the plane of the quark coupling g_B versus mass $m_{Z'_B}$ for the baryonic Z' model described in Sect. 2.3.1. Quark couplings above the contours are excluded at 95% C.L. This plot is adapted from [71, 72]

resonances in the full dijet invariant mass distribution, including both at high and low masses, is interesting.

Dijet resonance searches have a long history at hadron colliders, for a review see [11]. Searches were conducted by the UA1 and UA2 experiments at CERN during the 1980s and 1990s in $\sqrt{s} = 0.63$ TeV $p\bar{p}$ collisions [47–52], the CDF and D0 experiments at Fermilab in the 1990s and 2000s in $\sqrt{s} = 1.8 - 1.96$ TeV $p\bar{p}$ collisions [53–59], and since 2010 by the ATLAS and CMS experiments at CERN in $\sqrt{s} = 7 - 8$ TeV pp collisions [60–70]. All of the dijet resonance searches performed so far have seen no evidence for new resonances. However, this should not deter experiments from continuing to search for dijet resonances. With each increase in centre-of-mass energy of a collider, higher dijet masses than ever before become accessible. As dataset sizes increase so does the sensitivity to small cross-section resonances. The centre-of-mass energy and dataset size are the key factors that determine the sensitivity of the search, and both have increased significantly over time.

By comparing the exclusion limits obtained from dijet resonance searches performed at experiments with different centre-of-mass energies and dataset sizes, we can see the impact these factors have on the sensitivity of the search. Figure 2.10 shows several exclusion limits in the plane of quark coupling g_B versus mass $m_{Z'_B}$ for the baryonic Z' model. It can be seen that the more recent experiments, which have a significantly higher centre-of-mass energy, extend the limits to much higher masses than previous experiments. This motivates performing the high mass dijet resonance search presented in this thesis since the LHC nearly doubled its centre-of-mass energy from $\sqrt{s} = 8$ TeV to $\sqrt{s} = 13$ TeV in 2015.

Figure 2.10[1] also illustrates the need to continue searching at low dijet invariant masses, as it shows that at low masses, in particular below 500 GeV, the limits are weaker, and some of the older experiments are setting the most stringent limits, even with small dataset sizes. The reason for this is that as experiments achieve higher centre-of-mass energies, the QCD background at low masses becomes immense [38]. In order to cope with the huge background rates, experiments are forced to introduce minimum thresholds on the transverse energy (E_T) of jets at the trigger level, and to discard increasing fractions of events if these thresholds are lower, resulting in a loss of sensitivity at low masses. In order to avoid these experimental restrictions, modern experiments must use new techniques in order to be sensitive to the lower dijet mass regions. This thesis presents two searches in which an initial state radiation object (a jet or a photon) is used to trigger the event; this reduces the E_T requirements on the jets that form the dijet, enabling us to efficiently gather low dijet mass events.

References

1. Weinberg S (2004) The making of the standard model. Eur Phys J C34:5–13. https://doi.org/10.1140/epjc/s2004-01761-1, arXiv:hep-ph/0401010 [hep-ph]
2. Novaes SF (1999) Standard model: an introduction. In: Particles and fields. Proceedings, 10th Jorge Andre Swieca summer school, Sao Paulo, Brazil, 6–12 Feb 1999. pp 5–102. arXiv:hep-ph/0001283 [hep-ph]
3. Pich A (2007) The Standard model of electroweak interactions. In: High-energy physics. Proceedings, European School, Aronsborg, Sweden, June 18–July 1 2006. pp 1–49. arXiv:0705.4264 [hep-ph]
4. Purcell A (2012) Go on a particle quest at the first CERN webfest. Le premier webfest du CERN se lance à la conquiête des particules. BUL-NA-2012-269
5. Higgs PW (1964) Broken symmetries and the masses of Gauge Bosons. Phys Rev Lett 13:508–509. https://doi.org/10.1103/PhysRevLett.13.508
6. Englert F, Brout R (1964) Broken symmetry and the mass of Gauge vector mesons. Phys Rev Lett 13:321–323. https://doi.org/10.1103/PhysRevLett.13.321
7. Khodjamirian A (2004) Quantum chromodynamics and hadrons: an elementary introduction. In: High-energy physics. Proceedings, European School, Tsakhkadzor, Armenia, 24 Aug–6 Sept 2003. pp 173–222. arXiv:hep-ph/0403145 [hep-ph]
8. Particle Data Group, Patrignani C et al (2016) Review of particle physics. Chin Phys C40.10. https://doi.org/10.1088/1674-1137/40/10/100001
9. Sarkar U (2008) Particle and astroparticle physics. Series in high energy physics, cosmology, and gravitation. Taylor & Francis, New York. https://doi.org/10.1201/9781584889328, ISBN: 9781584889311
10. Van J (1998) '98 electroweak interactions and unified theories. Moriond particle physics meetings. Ed. Frontières, ISBN: 9782863322444
11. Harris RM, Kousouris K (2011) Searches for dijet resonances at hadron colliders. Intl J Mod Phys A 26.30n31:5005–5055. https://doi.org/10.1142/S0217751X11054905, arXiv:1110.5302 [hep-ex]

[1] The majority of the limit contours were produced by reinterpreting model-independent limits, using the technique outlined by Dobrescu and Yu in [38]. The results published by the ATLAS Collaboration with a 20.3 fb^{-1} dataset collected at $\sqrt{s} = 8$ TeV [70] were added using the same technique. The CMS Scouting result was added by digitising the limit contour in [73] using WebPlotDigitizer [74].

12. Sjostrand T (2006) Monte Carlo generators. In: High-energy physics. Proceedings, European School, Aronsborg, Sweden, 18 June–1 July 2006. pp 51–74. arXiv:hep-ph/0611247 [hep-ph]
13. Gieseke S (2002) Event generators: new developments. In: Hadron collider physics. Proceedings, 14th Topical Conference, HCP 2002, Karlsruhe, Germany, 29 Sept–4 Oct 2002. pp 439–452. arXiv:hep-ph/0210294 [hep-ph]
14. Höche S (2014) Introduction to parton-shower event generators. In: Theoretical advanced study institute in elementary particle physics: journeys through the precision frontier: amplitudes for colliders (TASI 2014) Boulder, Colorado, 2–27 June 2014. arXiv:1411.4085 [hep-ph]
15. Alwall J (2007) An improved description of charged higgs boson production. Subnucl. Ser. 42:346–355. https://doi.org/10.1142/9789812708427_0012, arXiv:hep-ph/0410151 [hep-ph]
16. Collins JC, Soper DE, Sterman GF (1989) Factorization of hard processes in QCD. Adv Ser Direct High Energy Phys 5:1–91 https://doi.org/10.1142/9789814503266_0001. arXiv:hep-ph/0409313 [hep-ph]
17. Kumericki K (2016) Feynman diagrams for beginners. arXiv:1602.04182 [physics.ed-ph]
18. Reina L (2006) Lecture notes from 'Practical next-to-leading order calculation' course. CTEQ Summer School. http://www.hep.fsu.edu/~reina/talks/cteq06nlo.pdf
19. Tricoli A (2009) Underlying event studies at ATLAS. Technical report, ATL-PHYS-PROC-2009-048. Geneva: CERN
20. Andersson B et al (1983) Parton fragmentation and string dynamics. Phys Rept 97:31–145. https://doi.org/10.1016/0370-1573(83)90080-7
21. Webber BR (1984) A QCD model for jet fragmentation including soft gluon interference. Nucl Phys B 238:492–528. https://doi.org/10.1016/0550-3213(84)90333-X
22. Sjostrand T, Mrenna S, Skands PZ (2008) A brief introduction to PYTHIA 8.1. Comput Phys Commun 178:852–867. https://doi.org/10.1016/j.cpc.2008.01.036, arXiv:0710.3820 [hep-ph]
23. Gleisberg T et al (2009) Event generation with SHERPA 1.1. JHEP 02:007. https://doi.org/10.1088/1126-6708/2009/02/007, arXiv:0811.4622 [hep-ph]
24. GEANT4 Collaboration, Agostinelli S et al (2003) GEANT4: a simulation toolkit. Nucl Instrum Meth A506:250–303. https://doi.org/10.1016/S0168-9002(03)01368-8
25. ATLAS Collaboration (2010) The ATLAS simulation infrastructure. Eur Phys J C70:823–874. https://doi.org/10.1140/epjc/s10052-010-1429-9, arXiv:1005.4568 [physics.ins-det]
26. Webb S (2004) Out of this world: colliding universes, branes, strings, and other wild ideas of modern physics. Copernicus Series. Springer, Berlin. http://www.springer.com/gb/book/9780387029306, ISBN: 9780387029306
27. van den Bergh S (1999) The early history of dark matter. Publ Astron Soc Pac 111:657. https://doi.org/10.1086/316369, arXiv:astro-ph/9904251 [astro-ph]
28. Kunze KE (2016) An introduction to cosmology. In: Proceedings, 8th CERN-Latin-American school of high-energy physics (CLASHEP2015): Ibarra, Ecuador, 05–17 Mar 2015. pp 177–212. arXiv:1604.07817 [astro-ph.CO]
29. Zwicky F (1933) Die Rotverschiebung von extragalaktischen Nebeln. Helv Phys Acta 6:110–127. https://doi.org/10.1007/s10714-008-0707-4
30. Rubin VC, Thonnard N, Ford WK Jr (1980) Rotational properties of 21 SC galaxies with a large range of luminosities and radii, from NGC 4605 /R = 4kpc/ to UGC 2885 /R = 122 kpc/. Astrophys J 238:471. https://doi.org/10.1086/158003
31. Council NR, Sciences DEP, Astronomy BP, Universe CP (2003) Connecting quarks with the cosmos: eleven science questions for the new century. National Academies, Washington. https://www.nap.edu/catalog/10079/connecting-quarks-with-the-cosmos-eleven-science-questions-for-the, ISBN: 9780309171137
32. ESA/Hubble, NASA (2012) Hubble spies a spiral galaxy edge-on. https://www.nasa.gov/multimedia/imagegallery/image_feature_2210.html
33. NASA's Chandra X-ray Observatory (2016) NASA finds direct proof of dark matter. http://chandra.harvard.edu/press/06_releases/press_082106.html
34. Collaboration Planck (2016) Planck 2015 results. XIII. Cosmological parameters. Astron Astrophys 594:A13. https://doi.org/10.1051/0004-6361/201525830, arXiv:1502.01589 [astro-ph.CO]

35. Gelmini GB (2014) TASI 2014 lectures: the hunt for dark matter. In: Theoretical advanced study institute in elementary particle physics: journeys through the precision frontier: amplitudes for colliders (TASI 2014) Boulder, Colorado, 2–27 June 2014. arXiv:1502.01320 [hep-ph]
36. Chala M et al (2015) Constraining dark sectors with monojets and dijets. JHEP 07:089. https://doi.org/10.1007/JHEP07(2015)089, arXiv:1503.05916 [hep-ph]
37. Abercrombie D et al (2015) In: Boveia A et al (ed) Dark matter benchmark models for early LHC Run-2 searches: report of the ATLAS/CMS dark matter forum. arXiv:1507.00966 [hep-ex]
38. Dobrescu BA, Yu F (2013) Coupling-mass mapping of dijet peak searches. Phys Rev D 88:035021. https://doi.org/10.1103/PhysRevD.88.035021. arXiv:1306.2629 [hep-ph]
39. Altarelli G, Mele B, Ruiz-Altaba M (1989) Searching for new heavy vector bosons in $p\bar{p}$ Colliders. Z Phys C45:109. https://doi.org/10.1007/BF01556677. Erratum: Z Phys C47:676 (1990). https://doi.org/10.1007/BF01552335
40. Baur U, Hinchliffe I, Zeppenfeld D (1987) Excited quark production at hadron colliders. Intl J Mod Phys A 2:1285. https://doi.org/10.1142/S0217751X87000661
41. Baur U, Spira M, Zerwas PM (1990) Excited quark and lepton production at hadron colliders. Phys Rev D 42:815–824. https://doi.org/10.1103/PhysRevD.42.815
42. Arkani-Hamed N, Dimopoulos S, Dvali GR (1998) The hierarchy problem and new dimensions at a millimeter. Phys Lett B429:263–272. https://doi.org/10.1016/S0370-2693(98)00466-3, arXiv:hep-ph/9803315 [hep-ph]
43. Antoniadis I et al (1998) New dimensions at a millimeter to a Fermi and superstrings at a TeV. Phys Lett B436:257–263. https://doi.org/10.1016/S0370-2693(98)00860-0, arXiv:hep-ph/9804398 [hep-ph]
44. Randall L, Sundrum R (1999) A large mass hierarchy from a small extra dimension. Phys Rev Lett 83:3370–3373. https://doi.org/10.1103/PhysRevLett.83.3370, arXiv:hep-ph/9905221 [hep-ph]
45. Calmet X (ed) (2015) Quantum aspects of black holes. Fundam Theor Phys 178. https://doi.org/10.1007/978-3-319-10852-0
46. Meade P, Randall L (2008) Black holes and quantum gravity at the LHC. JHEP 05:003. https://doi.org/10.1088/1126-6708/2008/05/003, arXiv:0708.3017 [hep-ph]
47. UA1 Collaboration (1984) Angular distributions and structure functions from two jet events at the CERN SPS p anti-p collider. Phys Lett B 136:294. https://doi.org/10.1016/0370-2693(84)91164-X
48. UA1 Collaboration (1986) Measurement of the inclusive jet cross section at the CERN pp collider. In: Phys Lett B 172.3:461–466. https://doi.org/10.1016/0370-2693(86)90290-X, ISSN: 0370-2693
49. UA1 Collaboration (1988) Two jet mass distributions at the CERN proton-anti-proton collider. Phys Lett B 209:127–134. https://doi.org/10.1016/0370-2693(88)91843-6
50. UA2 Collaboration (1984) Measurement of jet production properties at the CERN Collider. Phys Lett B 144:283–290. https://doi.org/10.1016/0370-2693(84)91822-7
51. UA2 Collaboration (1991) A measurement of two-jet decays of the W and Z bosons at the CERN $\bar{p}p$ collider. Zeitschrift für Phys C Part Fields 49.1:17–28. https://doi.org/10.1007/BF01570793, ISSN: 1431-5858
52. UA2 Collaboration (1993) A search for new intermediate vector bosons and excited quarks decaying to two-jets at the CERN $\bar{p}p$ collider. Nucl Phys B 400.1:3–22. https://doi.org/10.1016/0550-3213(93)90395-6, ISSN: 0550-3213
53. CDF Collaboration (1990) Two-jet invariant mass distribution at $\sqrt{s} = 1.8$ TeV. Phys Rev D 41:1722–1725. https://doi.org/10.1103/PhysRevD.41.1722
54. CDF Collaboration (1995) Search for new particles decaying to dijets in $p\bar{p}$ Collisions at $\sqrt{s} = 1.8$ TeV. Phys Rev Lett 74:3538–3543. https://doi.org/10.1103/PhysRevLett.74.3538
55. CDF Collaboration (1993) Search for quark compositeness, axigluons, and heavy particles using the dijet invariant mass spectrum observed in $p\bar{p}$ collisions. Phys Rev Lett 71:2542–2546. https://doi.org/10.1103/PhysRevLett.71.2542

56. CDF Collaboration (1997) Search for new particles decaying to dijets at CDF. Phys Rev D 55:R5263–R5268. https://doi.org/10.1103/PhysRevD.55.R5263
57. CDF Collaboration (2009) Search for new particles decaying into dijets in proton-antiproton collisions at $\sqrt{s} = 1.96$ TeV. Phys Rev D 79:112002. https://doi.org/10.1103/PhysRevD.79.112002, arXiv:0812.4036 [hep-ex]
58. DØ Collaboration (2004) Search for new particles in the two-jet decay channel with the DØ detector. Phys Rev D 69:111101. https://doi.org/10.1103/PhysRevD.69.111101
59. DØ Collaboration (2009) Measurement of dijet angular distributions at $\sqrt{s} = 1.96$ TeV and searches for quark compositeness and extra spatial dimensions. Phys Rev Lett 103:191803. https://doi.org/10.1103/PhysRevLett.103.191803, arXiv:0906.4819 [hep-ex]
60. ATLAS Collaboration (2010) Search for new particles in two-jet final states in 7 TeV proton-proton collisions with the ATLAS detector at the LHC. Phys Rev Lett 105:161801. https://doi.org/10.1103/PhysRevLett.105.161801, arXiv:1008.2461 [hep-ex]
61. ATLAS Collaboration (2011) Search for quark contact interactions in dijet angular distributions in pp collisions at $\sqrt{s} = 7$ TeV measured with the ATLAS detector. Phys Lett B 694:327. https://doi.org/10.1016/j.physletb.2010.10.021, arXiv:1009.5069 [hep-ex]
62. Collaboration CMS (2010) Search for dijet resonances in 7 TeV pp collisions at CMS. Phys Rev Lett 105:211801. https://doi.org/10.1103/PhysRevLett.105.211801
63. CMS Collaboration (2010) Search for quark compositeness with the dijet centrality ratio in 7 TeV pp collisions. Phys Rev Lett 105:262001. https://doi.org/10.1103/PhysRevLett.105.262001, arXiv:1010.4439 [hep-ex]
64. CMS Collaboration (2011) Measurement of dijet angular distributions and search for quark compositiveness in pp Collisions at $\sqrt{s} = 7$ TeV. Phys Rev Lett 106:201804. https://doi.org/10.1103/PhysRevLett.106.201804, arXiv:1102.2020 [hep-ex]
65. CMS Collaboration (2011) Search for resonances in the dijet mass spectrum from 7 TeV pp collisions at CMS. Phys Lett B 704:123. https://doi.org/10.1016/j.physletb.2011.09.015, arXiv:1107.4771 [hep-ex]
66. ATLAS Collaboration (2011) Search for new physics in dijet mass and angular distributions in pp collisions at $\sqrt{s} = 7$ TeV measured with the ATLAS detector. New J Phys 13:053044. https://doi.org/10.1088/1367-2630/13/5/053044, arXiv:1103.3864 [hep-ex]
67. ATLAS Collaboration (2012) Search for new physics in the dijet mass distribution using 1 fb^{-1} of pp collision data at $\sqrt{s} = 7$ TeV collected by the ATLAS detector. Phys Lett B 708:37–54. https://doi.org/10.1016/j.physletb.2012.01.035, arXiv:1108.6311 [hep-ex]
68. ATLAS Collaboration (2013) ATLAS search for new phenomena in dijet mass and angular distributions using pp collisions at $\sqrt{s} = 7$ TeV. JHEP 1301:029. https://doi.org/10.1007/JHEP01(2013)029, arXiv:1210.1718 [hep-ex]
69. CMS Collaboration (2013) Search for narrow resonances using the dijet mass spectrum in pp collisions at $\sqrt{s} =$ 8TeV. Phys Rev D 87:114015. https://doi.org/10.1103/PhysRevD.87.114015, arXiv:1302.4794 [hep-ex]
70. ATLAS Collaboration (2015) Search for new phenomena in the dijet mass distribution using pp collision data at $\sqrt{s} = 8$ TeV with the ATLAS detector. Phys Rev D 91:052007. https://doi.org/10.1103/PhysRevD.91.052007, arXiv:1407.1376 [hep-ex]
71. ATLAS Collaboration (2015) Baryonic Z' summary plot. https://atlas.web.cern.ch/Atlas/GROUPS/PHYSICS/PAPERS/EXOT-2013-11/figaux_10.png
72. Boveia A (2017) Private communication
73. CMS Collaboration (2016) Search for narrow resonances in dijet final states at $\sqrt{s} = 8$ TeV with the novel CMS technique of data scouting. Phys Rev Lett 117.3:031802. https://doi.org/10.1103/PhysRevLett.117.031802, arXiv:1604.08907 [hep-ex]
74. Rohatgi A (2017) WebPlotDigitizer - web based plot digitizer version 3.11. http://arohatgi.info/WebPlotDigitizer

Chapter 3
The ATLAS Experiment

In order to search beyond the Standard Model of particle physics, and to study the Standard Model itself, a particle detector is needed. The ATLAS (**A** **T**oroidal **LHC** **A**pparatu**S**) detector is a large general-purpose particle detector which records the proton-proton collisions produced by the Large Hadron Collider (LHC). The analyses presented in this thesis utilise data which is recorded by the ATLAS detector. A detailed description of the ATLAS detector and the LHC are provided in [1, 2], respectively, and a very brief summary will be provided in this chapter.

The Large Hadron Collider is introduced in Sect. 3.1, and the ATLAS detector is described in Sect. 3.2, with particular focus given to the calorimeter systems, as these are essential to the reconstruction and study of jets.

3.1 The Large Hadron Collider

The proton-proton collisions recorded by the ATLAS detector in 2015 and 2016 had an unprecedented centre-of-mass energy of $\sqrt{s} = 13$ TeV. In order to achieve such high energies, bunches of protons are accelerated to increasing energies by a chain of particle accelerators, as illustrated in Fig. 3.1. The final accelerator in the chain is the Large Hadron Collider, in which the bunches of protons are accelerated in opposite directions in two separate vacuum tubes. The LHC ring is 27 km in circumference, and approximately 100 m underground. At four points around the ring, the bunches of protons from each tube cross, and the collisions are recorded by particle detectors, as shown in Fig. 3.1. One of these collision points is inside the ATLAS detector.

In addition to the centre-of-mass energy, another key quantity for an accelerator is the *instantaneous luminosity* $\mathcal{L}_{inst}$ which is related to the number of events per unit time $\frac{\mathrm{d}N}{\mathrm{d}t}$ and to the cross-section for an event to occur σ by the following equation [4]:

$$\mathcal{L}_{inst}\sigma = \frac{\mathrm{d}N}{\mathrm{d}t}. \tag{3.1}$$

L. A. Beresford, *Searches for Dijet Resonances*, Springer Theses,
https://doi.org/10.1007/978-3-319-97520-7_3

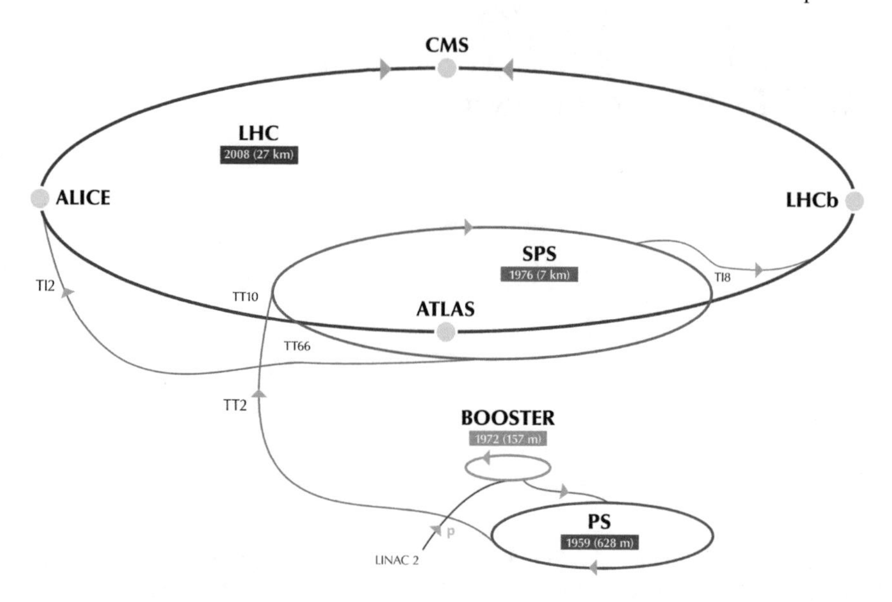

Fig. 3.1 This figure shows a diagram of the CERN accelerator complex. The speed and energy of the injected protons is increased by a chain of accelerators: a linear accelerator (Linac), the Booster, the Proton Synchrotron (PS), and the Super Proton Synchotron (SPS) accelerators, before entering the LHC ring. The protons are then accelerated further in opposite directions, before colliding inside the ALICE, CMS, LHCb and ATLAS detectors. This figure is adapted from [3]

By integrating the instantaneous luminosity with respect to time, we obtain the integrated luminosity $\mathcal{L}_{int}$ referred to as *luminosity* from now on. The luminosity is given by $\mathcal{L}_{int} = \frac{N}{\sigma}$, hence, in order to search for events with a low cross-section, a high luminosity is needed, making luminosity a very important quantity. The size of a dataset is quantified in terms of luminosity, with a larger dataset corresponding to a higher luminosity.

The luminosity gathered in a given time period is determined by the LHC running conditions and bunch parameters, for example, the number of protons in each bunch and the number of bunches [4]. When altering these parameters, there is a trade off between increasing the luminosity and the amount of pile-up, where pile-up refers to additional proton-proton interactions being included in the event. In the context of jets, these additional interactions can alter the energy of the jets in the event of interest, and they can also lead to additional jets being present. There are two types of pile-up, referred to as *in-time pile-up* and *out-of-time pile-up*. In-time pile-up occurs when multiple proton-proton interactions occur within one bunch-crossing [5]. In contrast, out-of-time pile-up occurs when signals from interactions in prior bunch-crossings are included in the event being processed [5], due to the energy measurement time being greater than the bunch spacing. In 2015, the number of bunches was increased and the bunch spacing was reduced from 50 to 25 ns, thus increasing the luminosity, but also increasing the out-of-time pile-up.

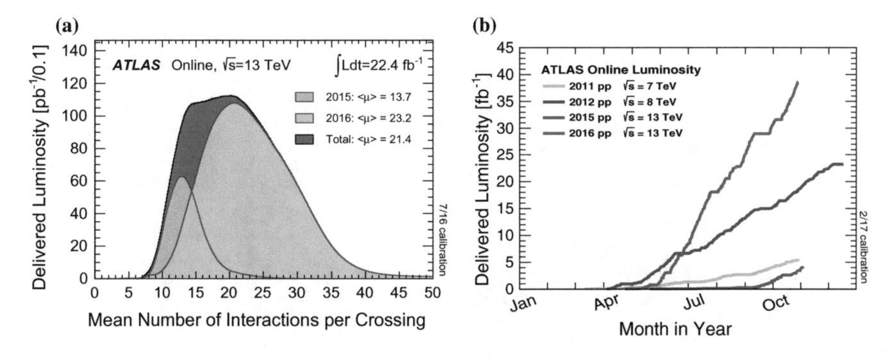

Fig. 3.2 **a**, taken from [8], shows the distribution of the mean number of inelastic proton-proton interactions per bunch crossing for data collected in 2015 and 2016. The value of $\sigma_{\text{inelastic}}$ in the calculation of μ is taken to be 80 mb. **b**, taken from [9], shows the luminosity as a function of time for the years 2011–2016

Two main variables are used to quantify the amount of pile-up for a given LHC running configuration. The first variable is the number of primary vertices in the event of interest N_{PV}, where a primary vertex corresponds to a proton-proton interaction point, and is the intersection point of at least two charged particle tracks [6]. This variable is used to quantify the amount of in-time pile-up. The second variable is the mean number of simultaneous inelastic proton-proton interactions being recorded in a single bunch crossing μ which is used to quantify the amount of out-of-time pile-up [5].

Figure 3.2a shows the distribution of μ for data collected in 2015 and 2016. On average, μ is higher for the data collected in 2016, indicating that there is more out-of-time pile-up. Figure 3.2b shows the integrated luminosity as a function of time for the years 2011–2016. The luminosity profile steeply rises in 2016, indicating that the running conditions and beam parameters have been altered to achieve a higher luminosity; for full details about the parameters utilised in 2015 and early 2016 see [7]. Note that the LHC operational period from 2009 to 2013 is referred to as Run I, and the operational period from 2015 to 2018 is referred to as Run II, with a two year upgrade period in between from 2013 to 2015.

In addition to categorising the proton-proton collision data recorded by ATLAS based on year, the data are further divided into *runs*. A run corresponds to a continuous period of data taking (typically a few hours long), during which many LHC and ATLAS configurations are fixed. For example, the proton-proton bunch spacing can not be changed during a run. Each run is further divided into luminosity blocks (typically one or two minutes of data taking), in which the instantaneous luminosity and the ATLAS and LHC conditions are assumed to be constant [10].

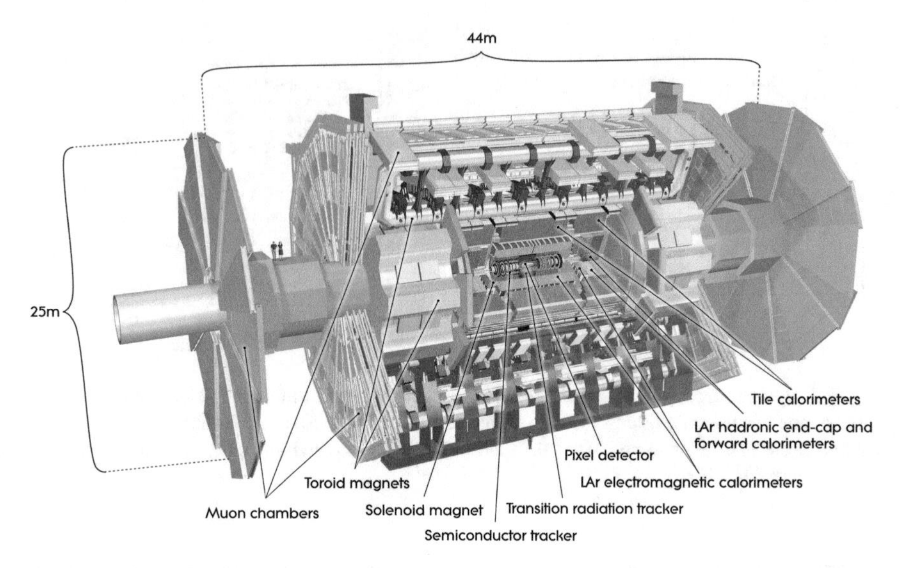

Fig. 3.3 The ATLAS detector and its sub-detectors are shown. This figure is taken from [12]

3.2 The ATLAS Detector

The ATLAS detector is cylindrical in shape, as shown in Fig. 3.3, and covers nearly the entire solid angle. The labels in Fig. 3.3 highlight the sub-detectors of which ATLAS is composed. The beam pipe passes through the centre of ATLAS, and is surrounded by layers of tracking detectors (collectively referred to as the inner detector) through which a 2 T solenoidal magnetic field passes. Outside of the inner detector are the electromagnetic and hadronic calorimeters, and surrounding the calorimeters is the muon spectrometer, which is immersed in a toroidal magnetic field. The ATLAS detector is divided into three regions: the barrel region in the centre, where sub-detectors form layers parallel to the beam pipe, the end-cap region, where the sub-detectors are positioned perpendicular to the beam pipe, and the forward region, where the calorimeters are positioned close to the beam line [11].

The ATLAS coordinate system is right-handed, with the origin at the centre of the ATLAS detector $(x, y, z) = (0, 0, 0)$, where the positive x direction points towards the centre of the LHC ring, the positive y direction points upwards, and the positive z direction points in the anti-clockwise direction, along the beam line. Often it is useful to use cylindrical coordinates, with the distance from the z-axis origin denoted r, and the azimuthal angle measured in the $x - y$ plane denoted ϕ. The rapidity is given by $y = \frac{1}{2}\ln\left(\frac{E+p_z}{E-p_z}\right)$, where E is the particle energy and p_z is the particle momentum in the $+z$ direction. Note that from now onwards, y will always refer to rapidity and not to the coordinate y, unless explicitly stated otherwise. For massless particles, the rapidity is equal to the pseudo-rapidity η, which is measured from the beam line, and is related to the polar angle θ by $\eta = -\ln(\tan(\frac{\theta}{2}))$. The transverse momentum of a

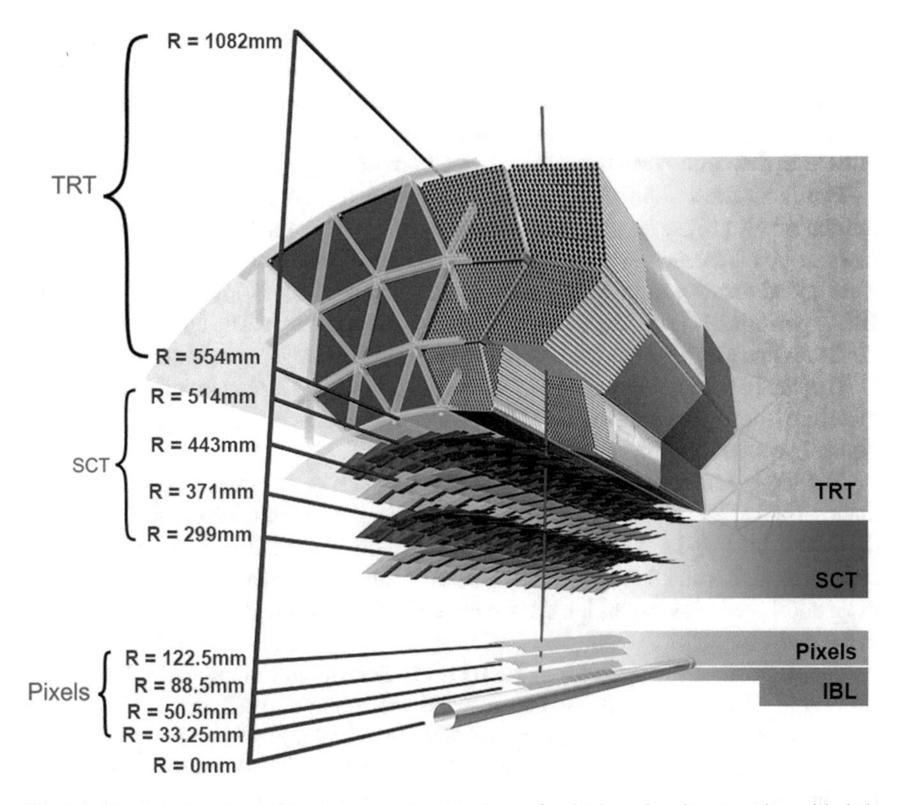

Fig. 3.4 The sub-detectors of the inner detector are shown for the barrel region, together with their distance from the beam line. From the beam line outwards, the sub-detectors are: the Insertable B-layer (IBL), the pixel detector, the Semiconductor Tracker (SCT), and the Transition Radiation Tracker (TRT). This figure is taken from [15]

particle p_T, and the transverse energy of a particle E_T, are measured in the $x - y$ plane. The angular separation between two particles in $\eta - \phi$ space is given by $\Delta\text{R} = \sqrt{(\Delta\eta)^2 + (\Delta\phi)^2}$. A brief description of each of the ATLAS sub-detectors will now be given, before describing the ATLAS trigger system.

3.2.1 Inner Detector

The Inner Detector (ID) provides precision tracking information for charged particles in the range $|\eta| < 2.5$. This is achieved through the use of fine granularity silicon detectors (pixels and micro-strips) [13] close to the beam pipe, surrounded by proportional drift tubes (straw tubes) filled with a Xenon or Argon gas mixture [14], as shown in Fig. 3.4.

Each of these detectors (pixel, strip and straw) record *hit* information, indicating the detection of a charged particle. The hits are then combined to reconstruct particle tracks, and subsequently vertices. Since the inner detector is immersed in a magnetic field, the trajectory of the charged particle is curved. The curvature of the track is utilised to determine the charge and momentum of the particle.

The inner-most sub-detector is the pixel detector, consisting of more than 80 million pixels [16], arranged into four layers of pixels in the barrel region and three disks of pixels in each end-cap region. The layer of pixels closest to the beam line were added for Run II, and are referred to as the Insertable B-layer (IBL) [16]. The IBL is positioned at a radius of 33.25 mm from the beam line. The close proximity to the beam line and high spatial resolution (design: $\sim$8 μm in $r - \phi$ and $\sim$40 μm in z [17]) of the IBL improves the precision of the tracking and vertexing information, and therefore improves the reconstruction of jets containing B-hadrons (b-jets), which typically contain a secondary vertex.

At a larger radius from the beam line, starting at $r = 299$ mm, lies the Semiconductor Tracker (SCT). The SCT consists of 4 layers of paired micro-strip detectors in the barrel, and 9 disks of paired micro-strip detectors in the end-cap. The micro-strips are paired with a small stereo angle between them, as this enables the measurement of an additional dimension, i.e. the z position in the barrel region, and the r position in the end-cap.

The outer-most sub-detector in the ID is the Transition Radiation Tracker (TRT), starting at 554 mm from the beam line and extending to 1082 mm, and covering the range $|\eta| < 2$. The TRT consists of more than 350,000 straw tubes [17] which are aligned parallel to the beam line in the barrel region, and arranged radially in the end-caps. These tubes provide $r - \phi$ hit position information in the barrel and $\phi - z$ information in the end-cap, and a typical charged particle will pass through at least 36 straws. In addition to providing tracking information, the TRT is also used for particle identification. The hits are classified into 'low threshold' hits and 'high threshold' hits, based on the size of the signal detected [18]. High threshold hits typically indicate the presence of transition radiation from the passage of an electron through the TRT.

3.2.2 Calorimeters

Outside the tracking system are the calorimeters. The purpose of calorimeters is to measure the energy of particles. In order to do this, layers of dense *absorber material* is used to induce a cascade of lower energy particles called a *shower*, and to contain the shower within the calorimeter. We can then reconstruct the energy of the particle which initiated the shower by measuring the energy of all the shower particles. The energy of the shower particles is measured in the *active material*, for example, through ionisation or scintillation. In ATLAS separate materials are utilised for the absorber and the active material, and these are arranged in alternating layers. This type of calorimeter is referred to as a *sampling calorimeter*. Details about shower development and calorimeter design are given in [19–21], and a brief overview will be provided here.

When high energy electrons, positrons or photons travel through the dense absorber material, energy losses due to bremsstrahlung and electron-positron pair production dominate. These processes produce a cascade of lower energy electrons, positrons and photons, collectively referred to as an *electromagnetic shower*. Once the particles produced in the shower reach a sufficiently low energy, ionisation interactions dominate and the particle shower ends.

The particle cascade produced by hadronic particles are referred to as *hadronic showers*. These are much more complex, due to the additional interactions via the strong force. As previously described, isolated partons hadronise creating sprays of particles called jets. Approximately 90% of the particles produced during hadronisation are pions (mesons composed of u and d quarks and anti-quarks) and $\sim\frac{1}{3}$ of these pions are neutral π^0. The decay of neutral pions typically results in the production of two photons, producing electromagnetic showers. The energy contained within the electromagnetic shower increases with the energy of the initial π^0, and hence, this contribution varies for each jet. In addition to electromagnetic interactions, there are also nuclear reactions. Many of these reactions produce charged particles which deposit energy via ionisation. However, some of the reactions result in the liberation of nucleons from nuclei, if the shower particle supplies enough energy to overcome the binding energy. The energy expended to overcome the binding energy is not recorded by the calorimeter, resulting in so-called *invisible energy*. Additionally, a small fraction of energy can be lost due to the production of neutrinos, which escape un-detected, resulting in *escaped energy*. Examples of these processes are shown in Fig. 3.5.

Due to the presence of invisible and escaped energy in hadronic showers, a calorimeter typically measures a lower fraction of the energy of the hadronic component of the shower, i.e. it has a lower *response* to the hadronic component than to the electromagnetic component. If this effect is not compensated for, then the calorimeter is described as *non-compensating*. The calorimeters employed by ATLAS are non-compensating, and hence, the mis-measurement of the hadronic energy needs to be accounted for by calibrations, as described in Chap. 4.

The shower profiles for electromagnetic and hadronic showers are conveniently expressed by the radiation length X_0, and the nuclear interaction length λ_{int}, respectively. The radiation length is the average length over which an electron loses $1 - \frac{1}{e} \sim 63\%$ of its energy via bremsstrahlung, and a photon travels on average $\frac{9}{7}X_0$ before undergoing pair production. A hadron, however, travels on average one λ_{int} before undergoing a nuclear reaction. Typically λ_{int} is much larger than X_0. As an example, for lead, which is a common absorber material, X_0 is 0.56 cm, and λ_{int} is 17.59 cm [20]. This indicates that generally more material is needed to contain the hadronic shower than the electromagnetic shower. This can be exploited for particle identification purposes, by having a separate *electromagnetic calorimeter* and *hadronic calorimeter*. The electromagnetic calorimeter aims to fully contain electromagnetic showers, and the hadronic calorimeter, which is positioned outside of the

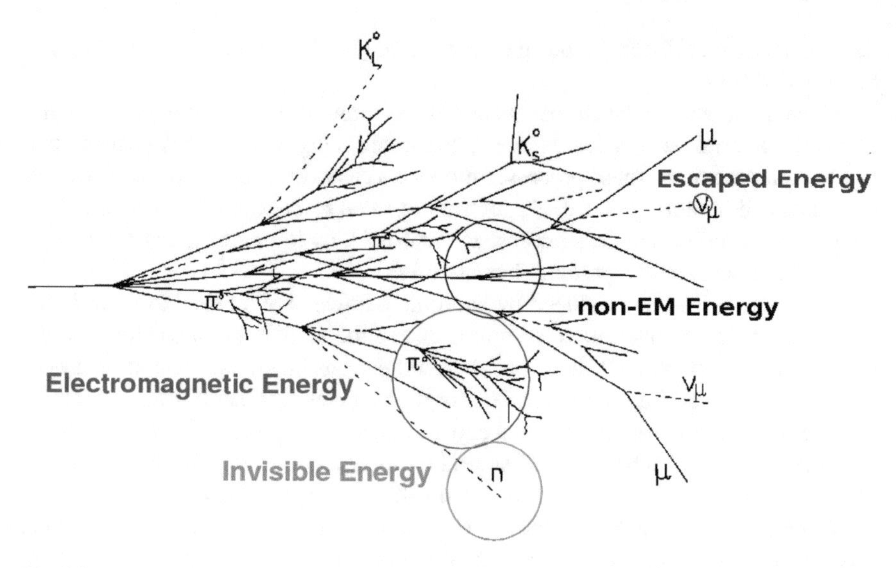

Fig. 3.5 A schematic diagram [22] of a hadronic shower, showing the contributions to the shower energy

electromagnetic calorimeter, aims to fully contain hadronic showers. Therefore, only hadronic particles should initiate showers which extend into the hadronic calorimeter. Note that both hadronic and electromagnetic interactions can occur in both detectors. Due to the larger depth needed to contain hadronic showers, it is not practical to build a hadronic calorimeter which can contain every particle created in the shower. Therefore, hadronic showers can 'leak' outside of the calorimeter. This effect is referred to as *longitudinal shower leakage* or *jet punch-through*.

The ATLAS calorimeter system is shown in Fig. 3.6. The electromagnetic calorimeter is more than 22 X_0 thick in the barrel region, and more than 24 X_0 thick in the end-cap region, so the electromagnetic shower should be well contained in the EM calorimeter. The hadronic calorimeter is approximately 9.7 λ_{int} in the barrel, and 10 λ_{int} in the end-caps; details about the material profile of the hadronic calorimeter will be given in Chap. 4, in the context of jet punch-through.

Electromagentic Calorimeter

The electromagnetic calorimeter (EM calorimeter) is a sampling calorimeter, utilising lead as the absorber and liquid Argon as the active material, covering the range $|\eta| < 3.2$. As shown in Fig. 3.6, the EM calorimeter is divided into the EM barrel calorimeter, spanning $|\eta| < 1.475$, and two EM end-cap calorimeters (EMEC), spanning $1.375 < |\eta| < 3.2$.

When a charged particle traverses the active layers of the calorimeter, it ionises the liquid Argon, liberating electrons. These electrons drift under the influence of an electric field and are collected on electrodes between the lead layers, producing a current proportional to the energy of the charged particle [24]. The lead absorber

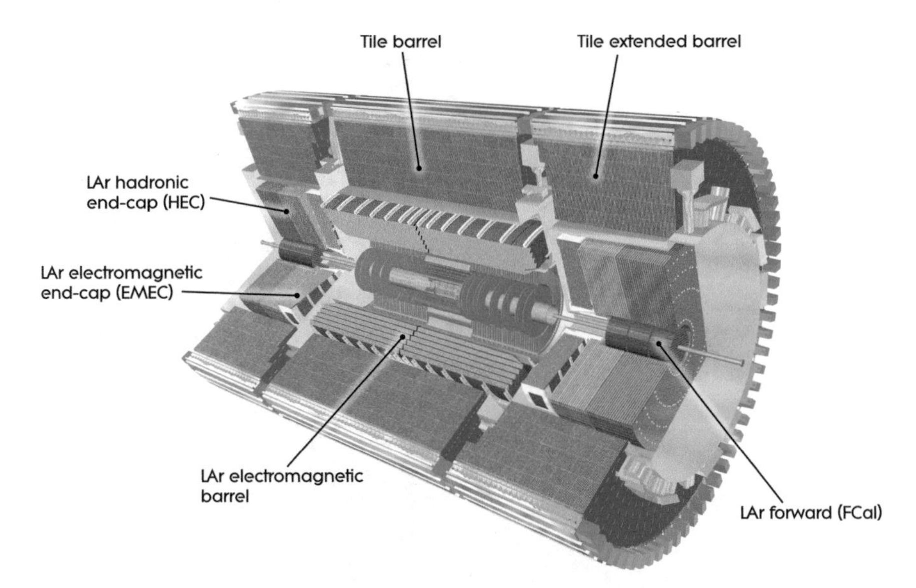

Fig. 3.6 The layout and sub-detectors of the calorimeter systems are shown. The calorimeters lie outside of the inner detector and the solenoidal magnet system. The electromagnetic calorimeter consists of the LAr barrel calorimeter, the inner LAr end-cap calorimeters, and the first layer of the forward calorimeters. The hadronic calorimeter consists of the tile barrel calorimeter, the tile extended barrel calorimeter, the outer LAr end-cap calorimeters, and the second two layers of the forward calorimeters. This figure is taken from [23]

layers and the electrodes are arranged in an accordion-shaped pattern, as shown in Fig. 3.7, ensuring full coverage in ϕ, without any cracks. The layers in the end-cap region also have an accordion-shaped pattern; however, they are arranged in the radial direction, rather than axially, as in the barrel region.

As shown in Fig. 3.7, the EM barrel calorimeter is divided up into three layers, referred to as layers 1, 2 and 3, or the front, middle and back layers, where layer 1 (the front layer) is closest to the beam line. The calorimeter is segmented in each layer, with the dimensions shown in the figure. Fine segmentation, i.e. small calorimeter cells, are utilised in the front layer, providing precise information about the shower position. The middle layer is coarser and records the majority of the electromagnetic shower energy, and the back layer is coarser still and records the tail of the electromagnetic shower. A single active layer of liquid Argon, referred to as a pre-sampler (PS), is placed before the EM calorimeter. The PS spans the region $|\eta| < 1.8$, and is used to estimate the energy lost by electrons, positrons and photons upstream of the EM calorimeter.

The region $1.37 < |\eta| < 1.52$ is referred to as the transition region or 'crack' region between the barrel and end-cap calorimeters. This region is poorly instrumented, due to the presence of cables and detector services, and the energy resolution (the spread of the measured energy with respect to the true value) is degraded. This region is usually excluded when utilising photons in an analysis, or when performing precise

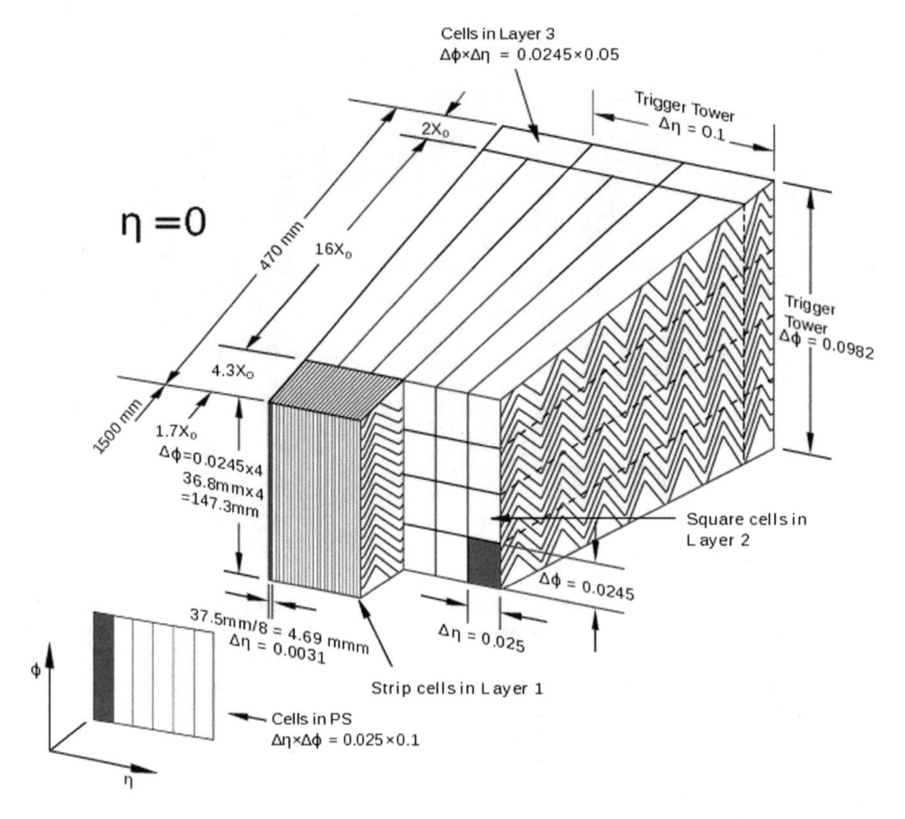

Fig. 3.7 This diagram shows the layout and segmentation of the electromagnetic barrel calorimeter and the pre-sampler (PS). The dimensions of the calorimeter cells and trigger towers are illustrated. This figure is taken from [25]

measurements involving electrons. Additionally, the region $|\eta| \geq 2.37$ is typically not included in analyses utilising photons as the fine granularity strips shown in Fig. 3.7 only extend to $|\eta| < 2.37$ [26].

Hadronic Calorimeter

Surrounding the EM calorimeters is the hadronic calorimeter. The hadronic calorimeter is divided into the tile calorimeter, two hadronic end-cap calorimeters, and two forward calorimeters. As shown in Fig. 3.6, the tile calorimeter is composed of the tile barrel calorimeter, spanning $|\eta| < 1$, and two tile extended barrel calorimeters, spanning $0.8 < |\eta| < 1.7$. The hadronic end-cap calorimeters (HEC) span $1.5 < |\eta| < 3.2$, and the forward calorimeter (FCal) spans $3.1 < |\eta| < 4.9$. Note that the transition regions between the detectors tend to be more poorly instrumented, with a reduction in material in these regions.

The tile calorimeter utilises alternating tiles of steel absorber and plastic scintillator as the active material, as shown in Fig. 3.8. The steel tiles are much thicker than the scintillator tiles, providing more material to help contain the shower. In both the

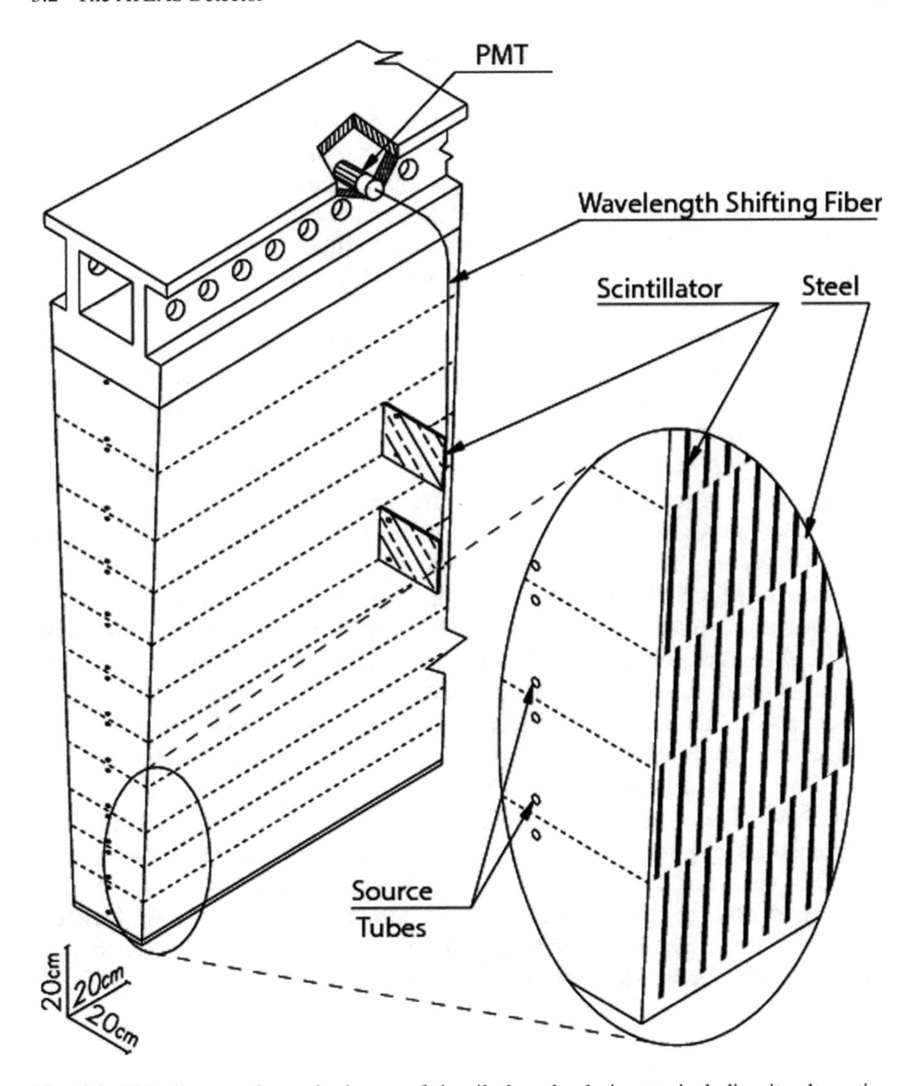

Fig. 3.8 This diagram shows the layout of the tile barrel calorimeter, including its alternating steel and scintillator structure and the light detection system. In the detector, this section would be positioned parallel to the beam line, with particles entering from the bottom of the section. This figure is taken from [29]

barrel and the extended barrel, the tiles are arranged in three layers, with a cell size of 0.1×0.1 ($\Delta\eta \times \Delta\phi$) in the first two layers and 0.2×0.1 ($\Delta\eta \times \Delta\phi$) in the final layer. This is coarser than the granularity in the EM calorimeter, as hadronic showers tend to be larger than electromagnetic showers, both laterally and longitudinally, and do not require such fine granularity. When a particle traverses the scintillator tiles, light is produced due to the excitation and de-excitation of atoms in the scintillator [27]. The light then passes through a wavelength shifting fiber, and is detected via a photo-multiplier tube (PMT), as illustrated in Fig. 3.8. The intensity of the detected light is proportional to the visible energy of the particle [28].

Due to the increased particle rates in the more forward η direction (mainly from small angle QCD scattering), the HEC calorimeter and the FCal need to be very radiation hard, and they need to be sufficiently dense to contain the showers in the limited space available. For both calorimeters, liquid Argon is used as the active material. Copper is used as the absorber for the HEC and for the first layer of the FCal (this layer primarily measures energy from electromagnetic interactions), and tungsten is used for the second two layers of the FCal (these two layers primarily measure the energy from hadronic interactions).

For each HEC calorimeter, flat copper plates forming two disks are utilised. For the FCal, a more complex geometry is utilised, consisting of an absorber matrix with cylindrical electrode tubes and rods, with a very narrow liquid Argon gap between the tube and the rod. Utilising a narrow liquid Argon gap ensures a fast readout time and reduces the buildup of ions, which can affect the electric field [30].

In order to obtain the energy deposited in each calorimeter cell (for both the EM and hadronic calorimeters), calibration constants are applied to the recorded signal in each cell [31]. The energy is then scaled to compensate for the energy deposited in the absorber layers [32]. We say that the energy has been calibrated to the *electromagnetic scale* (EM scale), as this is the correct cell energy for EM sources. However, it is an underestimate for hadronic sources, as we have not accounted for the invisible or escaped energy that is present in hadronic energy depositions. These factors are accounted for in the jet energy scale calibration.

3.2.3 Muon Spectrometer

Since muons typically do not deposit a large fraction of their energy in the calorimeter, they are able to traverse the calorimeter and reach the muon spectrometer. The muon spectrometer, illustrated in Fig. 3.9, provides precise momentum measurements for muons in the range $|\eta| < 2.7$. It also provides trigger capabilities in the range $|\eta| < 2.4$.

The measurement of muon momentum relies on the curvature of the muon tracks due to the large toroidal magnet system. This system provides a magnetic field which is typically perpendicular to the motion of the muon, with a field strength of $\sim$0.5 T in the barrel region and $\sim$1 T in the end-cap region. Precision tracking in the region $|\eta| < 2.7$ is achieved through the use of Monitored Drift Tubes (MDT) [34], noting

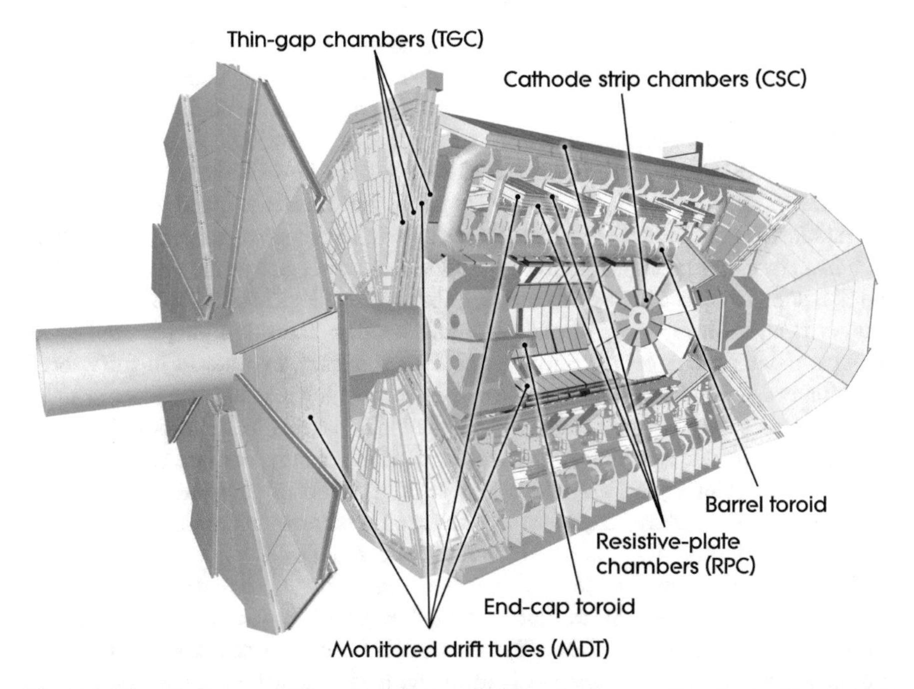

Fig. 3.9 The sub-detectors of the muon spectrometer are shown, as well as the toroidal magnet system. The Monitored Drift Tubes (MDT) and Cathode Strip Chambers (CSC) provide precision position information. The Resistive-plate Chambers (RPC) and Thin-gap Chambers (TGC) provide coarse position information and trigger capabilities. This figure is taken from [33]

that 'Monitored' does not relate to the type of drift tube, but to the fact that the tubes are monitored for mechanical deformations [35]. The MDTs are arranged into three layers in the barrel and four layers in the end-cap. The innermost layer of MDTs in the end-cap region is limited to $|\eta| < 2$, and a finer granularity sub-detector, the Cathode Strip Chamber (CSC) [36], is employed in the region $2 < |\eta| < 2.7$, due to the higher rate of particles in the more forward region.

Faster sub-detectors with coarser resolution are utilised for the trigger. These sub-detectors are the Resistive-plate Chambers (RPC) [37] for $|\eta| < 1.05$, and the Thin-gap Chambers (TGC) [38, 39] for $1.05 < |\eta| < 2.4$. In addition to being utilised for the trigger, these sub-detectors also provide additional ϕ position information.

3.2.4 *Trigger System*

Due to the extremely high collision rate at the LHC (40 MHz, 25 ns bunch spacing), it is not feasible to readout and store all of the data from the ATLAS sub-detectors for each of the collisions. Instead, the ATLAS *trigger system* is employed in order to decide which collisions are potentially interesting, i.e. those containing a certain number of physics objects, e.g. jets, or surpassing a threshold in energy. The trigger system has been updated for Run II, and a full description of the changes is provided in [40]; only a brief summary of the Run II trigger system will be provided here.

The ATLAS trigger system consists of two tiers: the Level 1 trigger (L1), and the High Level Trigger (HLT). The level 1 trigger is hardware-based, and is responsible for making fast decisions ($\sim$2.5 μs). This trigger utilises coarse granularity information from the calorimeter and muon systems to reduce the total event rate to 100 kHz. For the analyses described in this thesis, either a single jet trigger or a single photon trigger is utilised. In the L1 trigger a sliding window algorithm [41] is applied to groups of calorimeter cells, referred to as *trigger towers*, in order to identify local maxima in E_T. Regions of Interest (RoIs), corresponding to candidate jets or candidate photons/electrons, are defined around the local maxima. For photons/electrons, the EM calorimeter is scanned, and isolation criteria can be applied to reject candidates surrounded by high E_T towers in either the EM or hadronic calorimeters. For jets, both the EM and hadronic calorimeter are scanned. Note that the energies calculated in the L1 trigger are calibrated to the EM scale.

If an event is accepted by the L1 trigger, the event is then processed by the software-based HLT. At the HLT more time is available ($\mathcal{O}$ 200 ms) and finer granularity information is utilised, together with information from other detector sub-systems. For example, tracking information is available to distinguish between photon and electron candidates. In the HLT the physics objects are reconstructed and calibrated in a similar manner to the offline reconstruction and calibration. The HLT reduces the total event rate to 1 kHz. If the event passes the HLT trigger then the data for this potentially interesting event are read out and stored for offline reconstruction, calibration, and analysis. Note that if the HLT is unable to make a decision about a particular event in the assigned time, then the event is recorded to the *debug stream* [42]. Since the events in the debug stream could potentially be interesting, they are reprocessed offline and throughly investigated before inclusion in the analysis.

Due to the limited bandwidth available for all of the triggers utilised in ATLAS, for some particular triggers it is necessary to *prescale* them. This means that a certain fraction (equal to $\frac{1}{\text{prescale}}$) of the potentially interesting events selected by this trigger are rejected. Prescaling is needed to ensure that each trigger does not surpass their assigned bandwidth allowance. For single jet triggers, the lower the E_T threshold the higher the prescale, due to the increasing jet production cross-section at low E_T.

References

1. ATLAS Collaboration (2008) The ATLAS experiment at the CERN large hadron collider. J Instrum 3.08:S08003. https://doi.org/10.1088/1748-0221/3/08/S08003
2. Brüning OS et al (2004) LHC design report. CERN yellow reports: monographs. CERN, Geneva. https://cds.cern.ch/record/782076
3. Mobs E (2016) The CERN accelerator complex. Complexe des accélérateurs du CERN, General Photo. https://cds.cern.ch/record/2197559

4. Herr W, Muratori B (2006) Concept of luminosity. https://cds.cern.ch/record/941318
5. ATLAS Collaboration (2016) Performance of pile-up mitigation techniques for jets in pp collisions at $\sqrt{s} = 8\ TeV$ using the ATLAS detector. Eur Phys J C76.11:5811. https://doi.org/10.1140/epjc/s10052-016-4395-z, arXiv:1510.03823 [hep-ex]
6. ATLAS Collaboration, Meloni F (2016) Primary vertex reconstruction with the ATLAS detector. Technical report, ATL-PHYS-PROC-2016-163. Geneva: CERN
7. Eshraqi M, Trahern G (eds) (2016) LHC Run 2: results and challenges. In: Proceedings of 57th ICFA advanced beam dynamics workshop on high-intensity, High brightness and high power hadron beams (HB2016). Geneva, JACoW. http://accelconf.web.cern.ch/AccelConf/hb2016/papers/proceed.pdf, ISBN: 9783954501854
8. ATLAS Collaboration (2017) Luminosity public results Run 2. https://twiki.cern.ch/twiki/bin/view/AtlasPublic/LuminosityPublicResultsRun2
9. ATLAS Collaboration (2016) Data preparation public plots. https://atlas.web.cern.ch/Atlas/GROUPS/DATAPREPARATION/PublicPlots/2016/DataSummary/figs/intlumivsyear.eps
10. Buckingham RM et al (2011) Metadata aided run selection at ATLAS. J Phys Conf Ser 331.4:042030. https://doi.org/10.1088/1742-6596/331/4/042030
11. Dankers RJ (1997) The physics performance of and level 2 trigger for the inner detector of ATLAS (particle Detector, Muon Tracking, Cern). INSPIRE-888264. Ph.D. thesis. Twente U., Enschede, 1998
12. Pequenao J (2008) Computer generated image of the whole ATLAS detector. https://cds.cern.ch/record/1095924
13. Hartmann F (2012) Silicon tracking detectors in high-energy physics. Nucl Instrum Meth A666:25–46. https://doi.org/10.1016/j.nima.2011.11.005
14. Mindur B ((2016) ATLAS Transition Radiation Tracker (TRT): straw tubes for tracking and particle identification at the large hadron collider. Technical report, ATL-INDET-PROC-2016-001. Geneva: CERN
15. Potamianos K (2015) The upgraded pixel detector and the commissioning of the inner detector tracking of the ATLAS experiment for Run-2 at the large hadron collider. In: Proceedings, 2015 European physical society conference on high energy physics (EPS-HEP 2015), Vienna, Austria, 22–29 July 2015. p 261. arXiv:1608.07850 [physics.ins-det]
16. Capeans M et al (2010) ATLAS insertable B-layer technical design report. Technical report CERN-LHCC-2010-013. ATLAS-TDR-19
17. Butti P (2014) Advanced alignment of the ATLAS tracking system. Technical report, ATL-PHYSPROC- 2014-231. Geneva: CERN
18. ATLAS Collaboration, Hines E (2011) Performance of particle identification with the ATLAS transition radiation tracker. In: Particles and fields. Proceedings, meeting of the division of the American physical society, DPF 2011, Providence, USA, 9–13 Aug 2011. arXiv:1109.5925 [physics.ins-det]
19. Wigmans R (2008) Calorimetry. Sci Acta 2.1: 18. http://siba.unipv.it/fisica/ScientificaActa/volume_2_1/Wigmans.pdf
20. Particle Data Group, Patrignani C et al (2016) Review of particle physics. Chin Phys C40.10. https://doi.org/10.1088/1674-1137/40/10/100001
21. Fabjan CW, Gianotti F (2003) Calorimetry for particle physics. Rev Mod Phys 75:1243–1286. https://doi.org/10.1103/RevModPhys.75.1243
22. Grahn K-J (2009) A layer correlation technique for pion energy calibration at the 2004 ATLAS combined beam test. pp 751–757. https://doi.org/10.1109/NSSMIC.2009.5402211, arXiv:0911.2639 [physics.ins-det]
23. Pequenao J (2008) Computer generated image of the ATLAS calorimeter. https://cds.cern.ch/record/1095927
24. ATLAS Collaboration, Meng Z (2010) Performance of the ATLAS liquid argon calorimeter. In: Physics at the LHC2010. Proceedings, 5th Conference, PLHC2010, Hamburg, Germany, 7–12 June 2010. DESY-PROC-2010-01. pp 406–408
25. ATLAS Collaboration, Nikiforou N (2013) Performance of the ATLAS liquid argon calorimeter after three years of LHC operation and plans for a future upgrade. https://doi.org/10.1109/ANIMMA.2013.6728060. arXiv:1306.6756 [physics.ins-det]

26. Hance M (2012) Photon physics at the LHC: a measurement of inclusive isolated prompt photon production at $\sqrt{s} = 7$ TeV with the ATLAS detector. Springer Theses. Springer, Berlin. http://www.springer.com/gp/book/9783642330612, ISBN: 9783642330629
27. Mdhluli JE, Mellado B, Sideras-Haddad E (2017) Neutron irradiation and damage assessment of plastic scintillators of the Tile Calorimeter. J Phys: Conf Ser 802(1):012008. https://doi.org/10.1088/1742-6596/802/1/012008
28. Carrió F et al (2014) The sROD module for the ATLAS tile calorimeter Phase-II upgrade demonstrator. J Instrum 9(02):C02019. https://doi.org/10.1088/1748-0221/9/02/C02019
29. Sotto-Maior Peralva B (2013) Calibration and performance of the ATLAS tile calorimeter. In: Proceedings, international school on high energy physics: workshop on high energy physics in the near future. (LISHEP 2013), Rio de Janeiro, Brazil, 17–24 Mar 2013. arXiv:1305.0550 [physics.ins-det]
30. Artamonov A et al (2008) The ATLAS forward calorimeter. J Instrum 3(02):P02010. https://doi.org/10.1088/1748-0221/3/02/P02010
31. Aleksa M et al (2006) ATLAS combined testbeam: computation and validation of the electronic calibration constants for the electromagnetic calorimeter. Technical report, ATLLARG- PUB-2006-003. Geneva: CERN
32. ATLAS Collaboration (2014) Calorimeter calibration. https://twiki.cern.ch/twiki/bin/view/AtlasComputing/CalorimeterCalibration
33. Pequenao J (2008) Computer generated image of the ATLAS Muons subsystem. https://cds.cern.ch/record/1095929
34. ATLAS Muon Group (1994) Monitored drift tubes chambers for Muon spectroscopy in ATLAS. Technical report, ATL-MUON-94-044. ATL-M-PN-44. Geneva: CERN
35. Primor D et al (2007) A novel approach to track finding in a drift tube chamber. J Instrum 2(01):P01009. https://doi.org/10.1088/1748-0221/2/01/P01009
36. Argyropoulos T et al (2008) Cathode strip chambers in ATLAS: installation, commissioning and in situ performance. https://doi.org/10.1109/NSSMIC.2008.4774958
37. Cattani G, The RPC group (2011) The resistive plate chambers of the ATLAS experiment: performance studies. J Phys: Conf Ser 280(1):012001. https://doi.org/10.1088/1742-6596/280/1/012001
38. Nagai K (1996) Thin gap chambers in ATLAS. Nucl Instrum Meth A384:219–221. https://doi.org/10.1016/S0168-9002(96)01065-0
39. Majewski S et al (1983) A thin multiwire chamber operating in the high multiplication mode. Nucl Instrum Meth 217:265–271. https://doi.org/10.1016/0167-5087(83)90146-1
40. ATLAS Collaboration (2017) Performance of the ATLAS trigger system in 2015. Eur Phys J C77.5:317. https://doi.org/10.1140/epjc/s10052-017-4852-3, arXiv:1611.09661 [hep-ex]
41. Lampl W et al (2008) Calorimeter clustering algorithms: description and performance. Technical report, ATL-LARG-PUB-2008-002. Geneva: CERN
42. Bartsch V (2012) Experience with the custom-developed ATLAS offline trigger monitoring framework and reprocessing infrastructure. ATL-DAQ-PROC-2012-040

Chapter 4
Physics Object Reconstruction in ATLAS

Object reconstruction is a vital component of all analyses. It is the crucial step in which the electronic signals read out from the detector are combined to form objects which can be identified as particles. Once identified, the objects are then calibrated, such that their physical attributes (for example, their energy) are corrected for known detector effects. The calibrated objects can then be used in physics analyses.

This chapter outlines the reconstruction and calibration of the objects utilised in the analyses in this thesis. Section 4.1 provides a description of the reconstruction of jets, and Sect. 4.2 focuses on the steps involved in calibrating the reconstructed jets, and the associated uncertainty. Section 4.3 gives a detailed explanation of the derivation of the jet punch-through uncertainty, which is associated with the jet energy scale calibration, and is particularly important for high energy jets. In Sect. 4.4 the reconstruction of photons is detailed.

4.1 Jet Reconstruction

As previously mentioned, when a highly energetic parton is isolated, hadronisation occurs. The process of hadronisation results in a collimated shower of colourless particles, extending both laterally and longitudinally, as illustrated in Fig. 4.1. The goal of jet reconstruction is to cluster together the particles which were produced in the hadronisation of the initial parton, forming an object called a jet. The kinematics of the jet should represent the kinematics of the initial parton [1], allowing us to search for new particles which decay to partons by using jets. Experimentally, detector quantities must be used as an input to the jet reconstruction.

L. A. Beresford, *Searches for Dijet Resonances*, Springer Theses,
https://doi.org/10.1007/978-3-319-97520-7_4

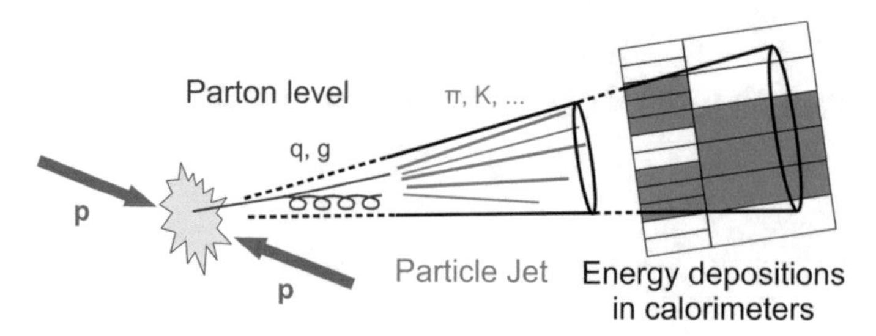

Fig. 4.1 This figure [2] illustrates the formation of jets. We start with the production of a parton in proton-proton collisions, which then hadronises to form a collimated spray of particles, a *particle jet*, which deposits energy in the calorimeters

4.1.1 Topological Clusters

In the analyses described in this thesis, *topological clusters* (topo-clusters) with a net positive energy are used as the input to the jet reconstruction [3]. An overview of topo-clusters and their construction is provided here, for further details see [3, 4]. A topo-cluster is a three dimensional collection of calorimeter cells. The calorimeter cells are grouped together based on their spatial separation, and their signal-to-noise ratio s_{cell}, where the signal is given by the absolute energy in the calorimeter cell, calibrated to the EM scale, and the noise is the quadrature sum of the average contributions from electronic noise and from pile-up. The aim of topological clustering is to capture the most significant regions, while reducing calorimeter noise and energy deposits from pile-up.

The construction of topo-clusters is illustrated in Fig. 4.2, and proceeds as follows: the first step is to categorise calorimeter cells based on their value of s_{cell}:

- **Primary seed cells**: $s_{\text{cell}} > 4$
- **Neighbour seed cells**: $s_{\text{cell}} > 2$
- **Border cell**: $s_{\text{cell}} > 0$

The primary seed cells are selected from the list of all calorimeter cells and are placed in order of decreasing s_{cell}. For each primary seed cell, starting with the cell with the highest s_{cell}, each of the neighbouring cells, both within the same sampling layer and in adjacent layers, are considered. The neighbouring cells are added to the topo-cluster if they are categorised as neighbour seed cells or border cells. Additionally, if a neighbour seed cell borders more than one topo-cluster, then the topo-clusters are merged. Once this procedure has been carried out for all of the primary seed cells, the process is repeated for the neighbour seed cells, until no seed cells remain. If the resulting topo-cluster displays more than one local signal maximum with a cell energy greater than 500 MeV, indicative of the topo-cluster containing the energy from more than one particle shower, then the topo-cluster is split. The use of topo-clusters as inputs to jet reconstruction has two primary benefits. Firstly, it reduces the number of inputs; if individual calorimeter cells were utilised, for example, then the

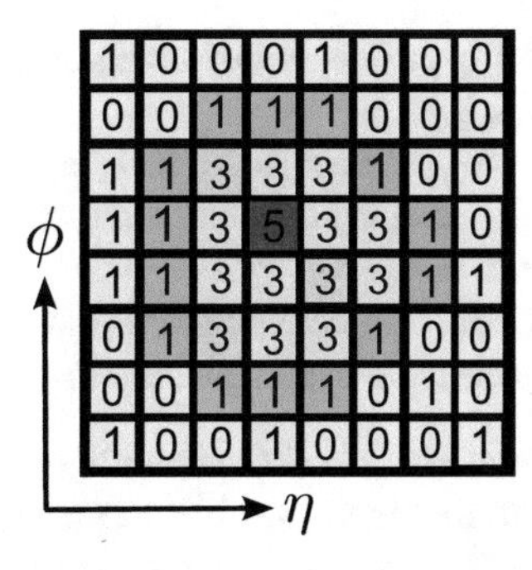

Fig. 4.2 This figure illustrates the construction of a topo-cluster from calorimeter cells. Primary seed cells (shown in red) have a signal-to-noise ratio > 4. Neighbour seed cells (shown in yellow) with a signal-to-noise ratio of > 2 are added to the topo-cluster. Finally, border cells (shown in green) are added to the topo-cluster. Figure adapted from [5]

jet reconstruction would proceed extremely slowly. Secondly, topo-clusters suppress extra energy added to the jet from noise and pile-up contributions.

For each topo-cluster, the energy E_{cluster} is calculated by summing the energy of its constituent cells, taking into account cases where the energy from cells was split between two topo-clusters in the cluster splitting step. The direction (η_{cluster}, ϕ_{cluster}) is calculated as the energy-weighted barycentre of the cell directions (η_{cell}, ϕ_{cell}), using the absolute value of the cell energies. The directions are calculated relative to the centre of the ATLAS detector $(x, y, z) = (0, 0, 0)$, where y refers to the coordinate not to rapidity. In the jet clustering algorithm, the topo-clusters are treated as massless and their transverse momentum p_T is calculated using E_{cluster}, η_{cluster} and ϕ_{cluster}.

4.1.2 Jet Clustering Algorithm

The most common jet clustering algorithm utilised in ATLAS is the anti-k_t jet clustering algorithm [6]. This algorithm is a sequential recombination algorithm, in which the inputs to the jet algorithm are clustered together based on a distance parameter $d_{i,j}$ between objects i and j. The term "object" refers to both inputs and to clustered groups of inputs. The distance parameter is given by the following equation:

$$d_{i,j} = \min\left(p_{T_i}^{-2}, p_{T_j}^{-2}\right)\frac{\Delta_{i,j}^2}{R^2}, \tag{4.1}$$

where $\Delta_{i,j}$ is the separation between objects i and j in the plane of rapidity y and azimuthal angle ϕ, given by $\Delta_{i,j}^2 = (y_i - y_j)^2 + (\phi_i - \phi_j)^2$, and R is the radius parameter, which is selected by the analyser. The clustering proceeds as follows:

1. The distance parameter $d_{i,j}$ and the object-beam distance $d_{i,B}$ are calculated for all input objects, where $d_{i,B} \equiv {p_{T_i}}^{-2}$.
2. If the smallest distance is $d_{i,j}$, then objects i and j are clustered together, forming a new object. If $d_{i,B}$ is the smallest distance, then object i is defined as a jet and is removed from the list of objects.
3. This process continues, with a re-calculation of the distances when new objects are formed, until all objects have been clustered into jets.

By inspecting Eq. (4.1) several features of the resulting jets can be identified. Firstly, we observe that $d_{i,j}$ is smaller for higher p_T objects, meaning that high p_T objects are clustered before lower p_T objects, resulting in jets centered around high p_T objects. Additionally, by taking the minimum ${p_T}^{-2}$ of the two objects, and therefore only taking into account the p_T of the highest p_T object in the pair i, j, emphasis is then placed on the geometrical requirement. In combination with the use of $d_{i,B}$, this means that two high p_T objects, 1 and 2, will only be clustered together if they are within a distance of $\Delta_{1,2} < R$. If objects 1 and 2 are within $R < \Delta_{i,j} < 2R$, then the higher p_T jet, jet 1 for example, will cluster together the lower p_T objects around it, forming a circular jet with radius R; jet 2, will then cluster together the lower p_T objects around it, forming a crescent shape. If objects 1 and 2 both have no higher p_T jets within $\Delta_{i,j} < 2R$ then they will each form circular jets of radius R. These features are illustrated in Fig. 4.3.

An important characteristic of the anti-k_t algorithm is that it can be applied to detector level quantities, hadrons, or partons in exactly the same way, aiding the comparison between experimental results and theoretical calculations [7]. In addi-

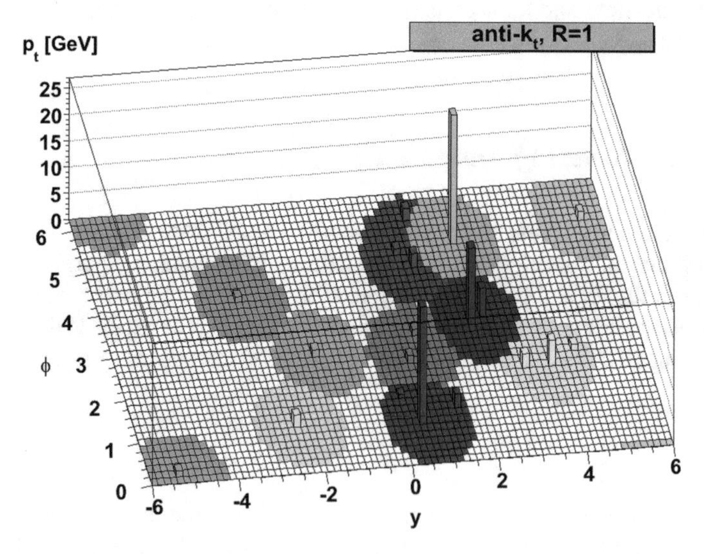

Fig. 4.3 This figure [6] shows the result of applying the anti-k_t algorithm to an event containing partons and $\sim 10^4$ very low p_T particles

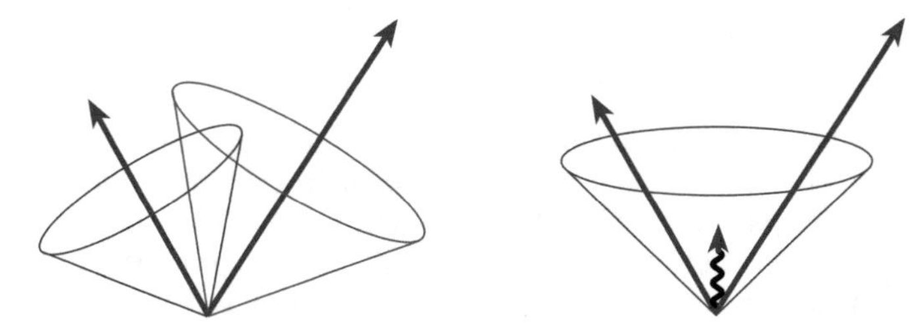

Fig. 4.4 This figure, taken from [1], illustrates a non-infrared safe scenario, in which the emission of a soft gluon has changed the outcome of the jet algorithm

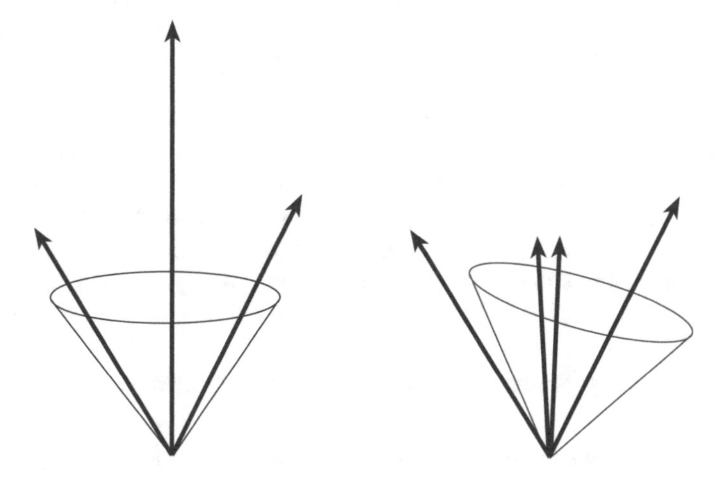

Fig. 4.5 This figure, taken from [1], illustrates a non-collinear safe scenario, in which the splitting of a parton into two nearly collinear partons has changed the outcome of the jet algorithm

tion, the jets constructed by the algorithm are both *infrared safe* and *collinear safe*. The terms infrared safe and collinear safe mean that the set of jets obtained by the algorithm should not be affected by the presence of soft (low p_T) emissions or by the splitting of a parton into two nearly collinear partons, respectively [8]. If these conditions were not met then experimental results could not be compared with theoretical calculations due to the presence of divergences in the calculations [7]. Illustration of a non-infrared safe scenario is provided in Fig. 4.4, and a non-collinear safe scenario is illustrated in Fig. 4.5.

The radius parameter R utilised in the analyses in this thesis is $R = 0.4$. This radius parameter is standard in ATLAS, and centrally derived calibrations and uncertainties are provided for jets with this radius parameter.

4.1.3 Reconstructed Jets

After the application of the jet algorithm, we have a set of jet objects. The next step is to choose a *recombination scheme*, i.e. a method for calculating their energy and momenta [1]. The *four-vector* recombination scheme is utilised, in which the four-vector for each jet is calculated by summing together the four-vectors of each of its constituent topo-clusters [9]. This scheme gives rise to a non-zero jet mass, despite the input topo-clusters being treated as massless.

The final step is to associate tracks from charged particles within the jet to the reconstructed jet. The *ghost association* technique [10] is used, in which the tracks are assigned infinitesimal p_T and are added to the list of jet inputs. The jet algorithm is then applied, and any tracks which are clustered into a jet are 'associated' to it.

Note that in reconstructed level Monte Carlo simulations topo-clusters are used as the inputs to the jet clustering algorithm. However, for truth level Monte Carlo, stable simulated particles with a lifetime, τ, satisfying $c\tau > 10$ mm are used as the inputs to the jet clustering algorithm, with the exception of neutrinos, muons and particles produced due to pile-up, which are not included [11]. Therefore, truth level MC jets are at the particle level, as illustrated in Fig. 4.1. The next step is to calibrate the four-momentum of the jets, such that they can be used in analyses.

4.2 Jet Energy Scale Calibration and Uncertainty

The calibration applied to the jets is referred to as the *Jet Energy Scale* (JES) calibration; however, most stages correct the full four-momentum of the jet, not just the jet energy. The goal of the calibration is to correct the jet four-momentum at the EM scale to the jet four-momentum at the particle level [12], as illustrated in Fig. 4.1. In order to do this, several detector effects must be corrected for, including: the lower response to hadronic showers (non-compensation) of the ATLAS calorimeter, energy deposited in dead material or before reaching the calorimeters, longitudinal shower leakage outside of the calorimeter (jet punch-through), energy losses due to noise thresholds or energy deposition outside of the jet, and additional energy due to pile-up [13]. Full details of the JES calibration and its corresponding uncertainty can be found in [11], and a summary of the relevant details will be provided here. The JES calibration involves multiple stages, as illustrated in Fig. 4.6.

For specific details about the JES calibration and uncertainty utilised in the high mass dijet analysis see [14]. The high mass dijet analysis was one of the first analyses to be performed using the $\sqrt{s} = 13$ TeV 2015 data. Hence, sufficient $\sqrt{s} = 13$ TeV data was not available in time to perform the residual in situ calibration. This calibration was instead derived using a combination of the Run I correction derived using $\sqrt{s} = 8$ TeV data, and Monte Carlo comparisons to quantify the impact of changes affecting simulation between Run I and Run II. The dijet + ISR analyses were performed later on, and hence, utilise the $\sqrt{s} = 13$ TeV data to derive the JES calibration and uncertainty.

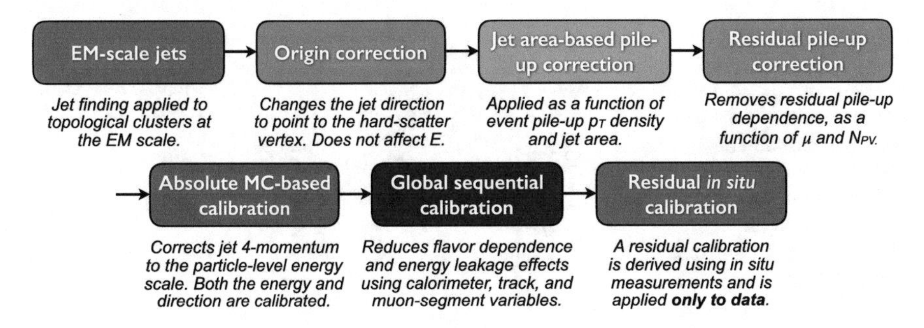

Fig. 4.6 This figure, adapted from [11], summarises the jet calibration chain used in ATLAS. For each stage, the name and a brief description of the calibration is given

4.2.1 Origin Correction

As previously mentioned, the direction of the topo-clusters, and subsequently the direction of the resulting jet is derived with respect to the centre of the ATLAS detector $(x, y, z) = (0, 0, 0)$, where y refers to the coordinate not to rapidity. The first step in the calibration is to adjust the jet four-momentum to point towards the primary vertex of the interaction (the vertex with the highest $\sum p_T^2$ of the associated tracks), improving the η resolution of the jet. This step does not alter the jet energy.

4.2.2 Pile-Up Corrections

The next stage is to remove pile-up contributions using two separate corrections, which are described fully in [15]. The first correction, referred to as the *jet area-based pile-up correction*, uses an estimate of the p_T density of the pile-up contribution for each event, calculated in the $|\eta| < 2$ region. This p_T density is then multiplied by the jet area,[1] proving a jet-level estimate of the pile-up contribution to the jet p_T, which is then subtracted from the measured jet p_T. Note that the pile-up p_T density estimate and the jet area calculation are both calculated in data when deriving the correction for data, and are both calculated in MC when deriving the correction for MC.

A second correction is then applied, which subtracts off any residual dependence of p_T on the number of primary vertices N_{PV} and the mean number of simultaneous inelastic proton-proton interactions being recorded in a single bunch crossing

[1]The jet area is calculated using ghost association, where a large number of 'ghost' particles with infinitessimal transverse momentum are added uniformly to the event in the $y - \phi$ plane. The number of these particles clustered to a jet gives a measure of the jet area [16].

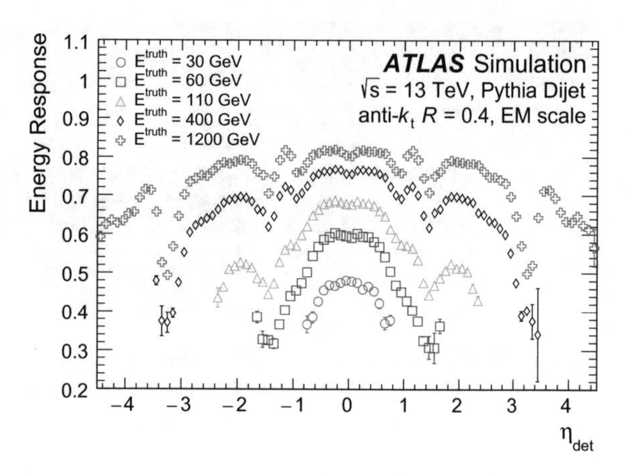

Fig. 4.7 The energy response is shown as a function of η_{detector}, before the application of the jet energy scale calibration. This illustrates that there are differences in the response of the detector to jets in different regions, and to jets with different energies. This figure is adapted from [11]

μ. The dependence is calculated in Monte Carlo as the difference in p_T between reconstructed level jets and their geometrically matched ($\Delta R < 0.3$) truth level jet, as truth level jets are not sensitive to pile-up. The ratio of the corrected p_T to the original p_T provides a scaling factor to correct the full four-momentum.

4.2.3 Absolute MC-Based Calibration

The next step in the sequence is to apply the *absolute MC-based calibration*. This step restores the jet four-momentum to the particle level (i.e. the level of the truth jets in Monte Carlo). In order to do this, geometrically matched reconstructed level and truth level jets are used to calculate the average energy response $\langle \frac{E_{\text{reco}}}{E_{\text{truth}}} \rangle$ as a function of E_{reco} in bins of η_{detector}. The *detector* η, η_{detector}, in which the jet direction is derived with respect to the centre of the ATLAS detector is utilised for the binning as this corresponds directly to the positions of the ATLAS sub-detectors. The correction is then given by the inverse of the obtained response function.

The average jet energy response as a function of η_{detector} before the application of the correction is shown in Fig. 4.7. The figure shows that the response is less than one, with a poorer response for jets with lower energy and for jets spanning the transition regions in the detector, i.e. between the barrel and end-cap system at $\sim\eta_{\text{detector}} = 1.4$, and between the end-cap and forward system at $\sim\eta_{\text{detector}} = 3.1$. After the application of this correction, differences in jet η_{reco} and jet η_{truth} are corrected for in a dedicated jet η calibration, altering the jet η and jet p_T, but not the jet energy.

4.2.4 Global Sequential Calibration

After restoring the jets to the particle level, the *global sequential calibration* is applied. This correction reduces the dependence of the response on selected variables, while maintaining the same average jet energy. This helps to equalise the response to jets with different properties, for example, between quark and gluon initiated jets. In addition, this correction reduces the *Jet Energy Resolution* (JER), i.e. the spread of the measured jet energies with respect to their true value. The following variables are utilised in the correction:

- **Calorimeter variables**: the fraction of the jet energy recorded in the first layer of the hadronic tile calorimeter (for $|\eta_{\text{detector}}| < 1.7$) and in the final layer of the EM calorimeter (for $|\eta_{\text{detector}}| < 3.5$). The calorimeter layer information characterises the energy profile of the jet, and jets which deposit large fractions of their energy in the tile calorimeter can be mis-measured due to the non-compensation of the ATLAS calorimeter.
- **Tracking variables**: the number of tracks and the jet width, i.e. the average transverse distance between the jet axis and tracks, weighted by the track p_T. Both of these variables utilise tracks with $p_T > 1\,\text{GeV}$ which are associated to the jet, and are in $|\eta_{\text{detector}}| < 2.5$. The tracking variables are used to distinguish between gluon-initiated jets and light quark-initiated jets. Gluon-initiated jets tend to result in wider jets containing more low p_T particles, to which the calorimeter typically has a lower response. Additionally, the jet width gives a measure of the number of particles in the jet which were measured in transition regions in the calorimeter.
- **Muon spectrometer variable**: muon segments are partial tracks in the muon spectrometer. They are ghost associated to jets in the same way as tracks are associated to jets. The number of muon segments associated to the jet N_{Segments} is an indicator of how well contained the jet is within the calorimeter. Highly energetic jets can longitudinally 'leak' outside of the calorimeter and interact in the muon spectrometer, meaning that the full hadronic shower is not captured by the calorimeter. If the jet is not fully contained then the jet energy is under-estimated, causing increased low jet energy response tails, which impacts the jet energy resolution. The longitudinal shower leakage (jet punch-through) is correlated with the number of muon segments associated to the jet, with a larger N_{Segments} indicating more longitudinal shower leakage (higher jet punch-through).

The correction is applied in a similar manner to the absolute MC-based calibration; however, the average transverse momentum response $\langle \frac{p_{T_{\text{reco}}}}{p_{T_{\text{truth}}}} \rangle$ is utilised, and is parametrised as a function of both $p_{T_{\text{reco}}}$ and the variable being corrected for, in bins of η_{detector}. The exception being the jet punch-through correction, in which the energy response is utilised, and is parametrised as a function of E_{reco} and $\log(N_{\text{Segments}})$. For full details about the jet punch-through correction see [17]. The inverse of the obtained response function provides the correction, and an additional constant ensures that the average jet energy is maintained. The correction is applied sequentially, with the dependence on one variable being corrected for, before the correction for the next variable is derived and applied.

Jet Punch-Through

Before showing the impact of the jet punch-through correction (the final stage of the global sequential calibration) a review of jet punch-through in Run II will be given; for a review of the jet punch-through properties studied in Run I see [18]. As previously described, the variable used to indicate that a shower is not fully contained inside the calorimeter is the number of muon segments associated to the jet N_{Segments}. Therefore, it is important to study the modelling of N_{Segments} in Monte Carlo. Note that the data and Monte Carlo samples utilised for all the punch-through studies shown in this chapter (with the exception of the jet punch-through correction) are the same as those used in the high mass dijet analysis described in Sect. 5.1.2. The analysis selection utilised for the data-MC comparison studies in this section is the same as the one applied in the derivation of the punch-through uncertainty, given in Table 4.1, with the exception of the $|\Delta\phi|$ and third jet p_T requirements, which are not applied.

Figure 4.8 shows the N_{Segments} distribution in data and MC for the two highest p_T jets. Note that in this thesis, the highest p_T jet is referred to as the *leading jet*, the second highest p_T jet is referred to as the *sub-leading jet*, and the two highest p_T jets are collectively referred to as the *leading jets*. The N_{Segments} distribution is shown to be well modelled by the Monte Carlo for lower N_{Segments} values. However, for higher N_{Segments} values, an excess of events in data are observed, with respect to Monte Carlo.

Table 4.1 The analysis selection used in the derivation of the punch-through uncertainty

Punch-through uncertainty selection
Remove events which:
–Do not belong to good LBs, defined in the good run list;
–Show evidence for noise bursts;
–Show evidence for data corruption in the calorimeters;
–Were recorded during the recovery procedure for the SCT;
–Are incomplete, i.e. they do not have information from the full detector.
Events must:
–Have a primary vertex with at least two associated tracks;
–Contain at least two clean jets with $p_T > 50\,\text{GeV}$;
–Pass HLT_j360 trigger;
–Contain a jet with $p_T \geq 410\,\text{GeV}$.
Two leading jets must satisfy:
–$|\eta_{\text{detector}}| < 2.7$;
–$|\Delta\phi| > 2.9$.
If there is a third jet present, it must be clean and satisfy $p_T < \max(12, 0.25\text{p}_\text{T}^{\text{average}})$.

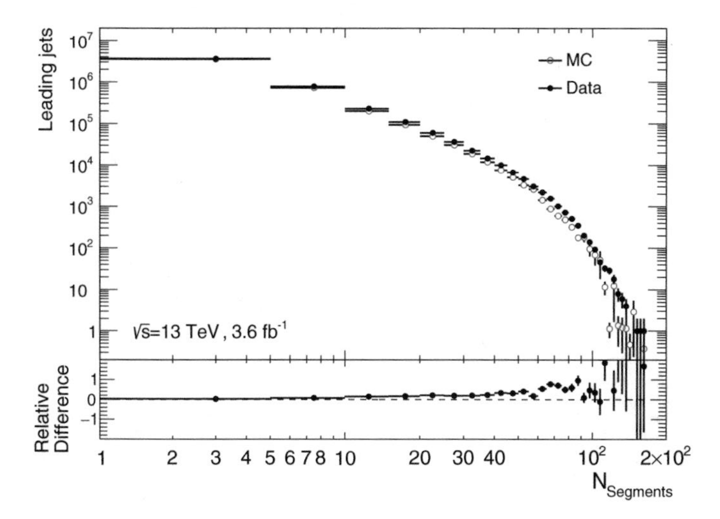

Fig. 4.8 A comparison between the N_{Segments} distribution for the leading two jets in data (shown in black), and Monte Carlo (shown in red). Note that the Monte Carlo histogram has been scaled to the integral of the data histogram. The N_{Segments} distribution is shown to be well modelled at low values of N_{Segments}. However, for higher N_{Segments} values an excess of events is observed in data, with respect to Monte Carlo

The cause of the mis-modelling of N_{Segments} is unknown; one suggestion from the ATLAS Simulation group is that the excess in data could be from cavern background (low energy particles, mainly photons and neutrons, filling the cavern during collision time) or other non-collision backgrounds producing hits in the muon spectrometer, which can form additional muon segments [19]. Cavern background is not included in standard Monte Carlo simulations. In Run I, improvements were observed in the modelling of N_{Segments} when using Monte Carlo samples with backgrounds measured in data overlaid [17]. However, such samples were not yet available for Run II Monte Carlo. Further studies would be necessary to confirm the source of the mis-modelling.

The amount of jet punch-through (indicated by N_{Segments}) is expected to increase with jet energy, and in regions of the detector with less material, i.e. corresponding to a low number of nuclear interaction lengths λ_I. The length of material (measured in λ_I) needed to contain 95% of the longitudinal component of the hadronic shower $L_{95\%}$ is related to the jet energy E by the equation

$$L_{95\%}[\lambda_I] = 0.6\ln(\text{E})[\text{GeV}] + 4\text{E}^{0.15} - 0.2, \tag{4.2}$$

taken from [20]. In Run I, N_{Segments} was observed to increase with jet energy, and to be enhanced in regions of the detector with less material [17]. These trends have also been studied using the Run II data and Monte Carlo, and are shown in Figs. 4.9 and 4.10a. In Fig. 4.9, N_{Segments} is shown as a function of jet energy for a restricted η_{detector} range to reduce the material dependence. N_{Segments} is shown to increase with jet energy and to be well modelled by the Monte Carlo.

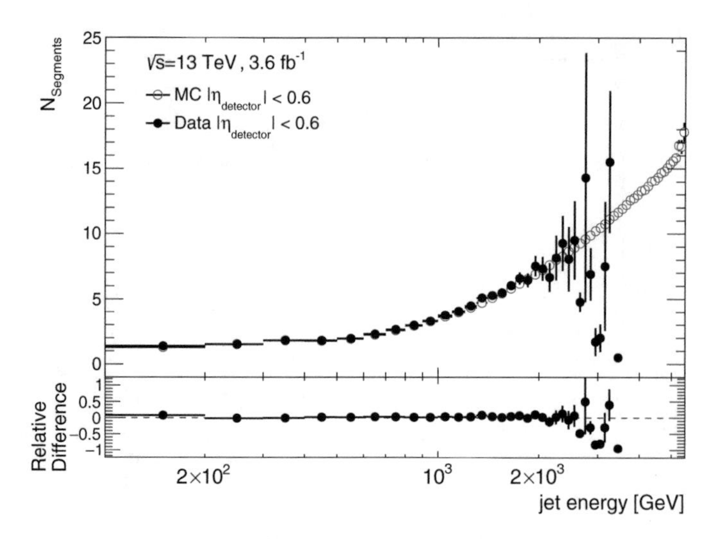

Fig. 4.9 The mean N_{Segments} is shown as a function of jet energy. Data is shown by the black points, and Monte Carlo is shown by the red points. N_{Segments} is seen to increase as jet energy increases

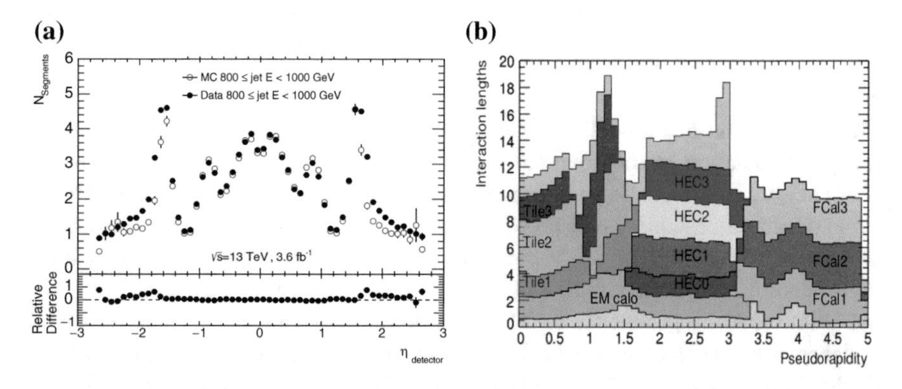

Fig. 4.10 In Figure **a** the mean N_{Segments} is shown as a function of η_{detector}. Data is shown by the black points, and Monte Carlo is shown by the red points. Figure **b** taken from [21], shows the amount of material (measured in units of interaction length) as a function of η_{detector}, with the total amount of material in front of the muon spectrometer shown in blue. By comparing Figure **a**, **b**, we see that the amount of material is anti-correlated with the observed N_{Segments}

In Fig. 4.10a, N_{Segments} is shown as a function of η_{detector} in a restricted jet energy range to reduce the jet energy dependence. For reference, the amount of material (measured in units of λ_I) versus η_{detector} for the ATLAS detector is provided in Fig. 4.10b. It is observed that for η_{detector} regions corresponding to a low amount of material we observe enhanced jet punch-through, i.e. high N_{Segments}, as expected. It is also observed that in central region, i.e. $|\eta_{\text{detector}}| < 1.6$, the Monte Carlo models the data well. However, in the more forward regions of the detector the modelling is poorer. The variation in the data-MC agreement with $\eta_{detector}$ justifies why the jet

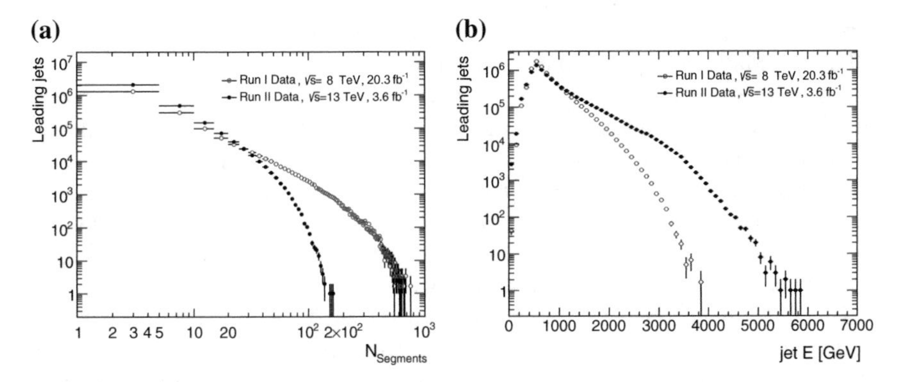

Fig. 4.11 A comparison between Run I data (20.3 fb^{-1} of $\sqrt{s} = 8$ TeV data, shown in red) and Run II data (3.6 fb^{-1} of $\sqrt{s} = 13$ TeV data, shown in black) for **a** the distribution of N_{Segments} for the leading two jets, and **b** the distribution of jet energies for the leading two jets. The comparison between the jet energy distributions shows that higher energies are reached with the Run II data; however, the comparison of the N_{Segments} distributions shows that the maximum N_{Segments} observed in Run II is much lower than the maximum N_{Segments} observed in Run I. Note that the Run I histograms have been scaled to the integral of the Run II histograms

punch-through calibration and uncertainty are derived in bins of η_{detector}, separating out each region and deriving a separate calibration factor and uncertainty. Note that by studying the η dependence of N_{Segments} in Run II Monte Carlo a bug causing jets with $|\eta| > 0.5$ to have no associated muon segments was identified and resolved prior to data taking, highlighting the importance of performing such studies.

Due to the relationship between N_{Segments} and jet energy, one would expect to observe an increase in the maximum N_{Segments} observed in Run II, with respect to the maximum N_{Segments} observed in Run I, since higher jet energies can be accessed due to the increase in the centre-of-mass energy. A comparison of the N_{Segments} distribution in Run I, and the N_{Segments} distribution in Run II was made, and the results are shown in Fig. 4.11a. Note that the leading jet p_T threshold in the event selection was increased to 460 GeV for the comparison, as this is the position from which the Run I lowest un-prescaled single jet trigger is 99.5% efficient. This figure shows that in fact the maximum N_{Segments} observed was higher in Run I than in Run II. By comparing the jet energy distribution in Run I and in Run II, shown in Fig. 4.11b, we verify that higher jet energies are indeed reached with the Run II data. This indicates that another factor must be responsible for the lower N_{Segments} reached in Run II. The decrease is attributed to the changes made to the muon segment reconstruction, and their association to jets between Run I and Run II. In the reconstruction of the muon segments, there has been a tightening of the timing thresholds in the muon spectrometer drift tubes, in order to reduce CPU consumption [14]. In the association of muon segments to jets, the ghost association technique is now in use. In Run I, the technique to associate muon segments to jets involved building muon segment containers during reconstruction, which included muon segments geometrically ΔR matched to anti-k_t R = 0.6 jets ($\Delta R < 0.4$). At the analysis level, the closest segment container to a jet (within $\Delta R < 0.3$) was associated to it. Both of these changes have led to the reduction in N_{Segments} observed in Run II.

An additional change between Run I and Run II is the introduction of a hit occupancy threshold in the muon segment reconstruction. The introduction of this threshold means that there could be a reduction in N_{Segments} when the threshold is exceeded. This is a problem as jets could falsely be assigned a lower number of N_{Segments}, due to the timing-out of the reconstruction. This effect will need to be studied in the future in order to assess the impact it has. However, at present this is not possible as there is no means to identify 'timed-out' events in data.

Due to the many changes between Run I and Run II, the jet punch-through correction was re-derived in Run II. Figure 4.12 shows the p_T response as a function of N_{Segments}, in bins of η_{detector} and $p_{T_{\text{truth}}}$, before and after the application of the jet punch-through correction, as well as the unit normalised N_{Segments} distribution below. The correction is shown to reduce the dependence of the p_T response on N_{Segments}. From these figures, it is seen that before the correction the response for the lower $p_{T_{\text{truth}}}$ bins is lower than for the higher $p_{T_{\text{truth}}}$ bins. This is counter-intuitive, since increased jet punch-through is expected for higher p_T jets. The reason for this observation is that at lower jet p_T, even though the absolute energy loss from jet punch-through is lower, the relative loss is higher, due to the lower p_T of the jet.

Note that, above $|\eta_{\text{detector}}| = 1.9$ there were insufficient events to derive the punch-through correction. Hence, the correction is not applied in this region, and bins corresponding to $|\eta_{\text{detector}}| = 1.9$ and above are not shown. Additionally, note that the response and N_{Segments} distribution is only shown for the region above $N_{\text{Segments}} = 20$. The punch-through correction is only applied to jets with at least 20 associated N_{Segments}. This threshold was chosen in order to avoid the over-correction of jets containing B-hadrons (b-jets), since the decay of a B-hadron leads to the production of at least one muon in the final state in ~20% of cases [23]. This muon creates a track in the muon spectrometer, and hence, creates several muon segments (typically 3 for a single muon), which could be misinterpreted as an indicator for jet punch-through.

The threshold in N_{Segments} above which the correction is applied was re-derived in Run II. This threshold is important as the lower the threshold is set, the more jets are corrected, increasing the impact of the jet punch-through correction. In order to decide where the threshold should be placed, the b-jet fraction as a function of N_{Segments}, shown in Fig. 4.13, was utilised. A jet is identified as a b-jet using the MV2c20 multivariate discriminant [24]. This algorithm utilises information such as vertex information and tracking information, to identify b-jets, exploiting features of B-hadron decays such as the presence of secondary vertices and tracks with large impact parameters with respect to the primary vertex. The 77% efficiency working point is used. Dijet events in which one of the jets (the *probe*) has $N_{\text{Segments}} \geq 1$, and the other jet (the *reference*) has $N_{\text{Segments}} = 0$ are utilised. The b-jet fraction plot is then produced by taking the ratio between the N_{Segments} distribution for all probe jets which are b-jets and the N_{Segments} distribution for all probe jets.

In Fig. 4.13 the fraction of b-jets is observed to rise and to fall off, plateauing after ~20 N_{Segments}. An increased b-jet fraction is observed for $N_{\text{Segments}} = 3$, which is the typical number of segments corresponding to a single muon track in the absence

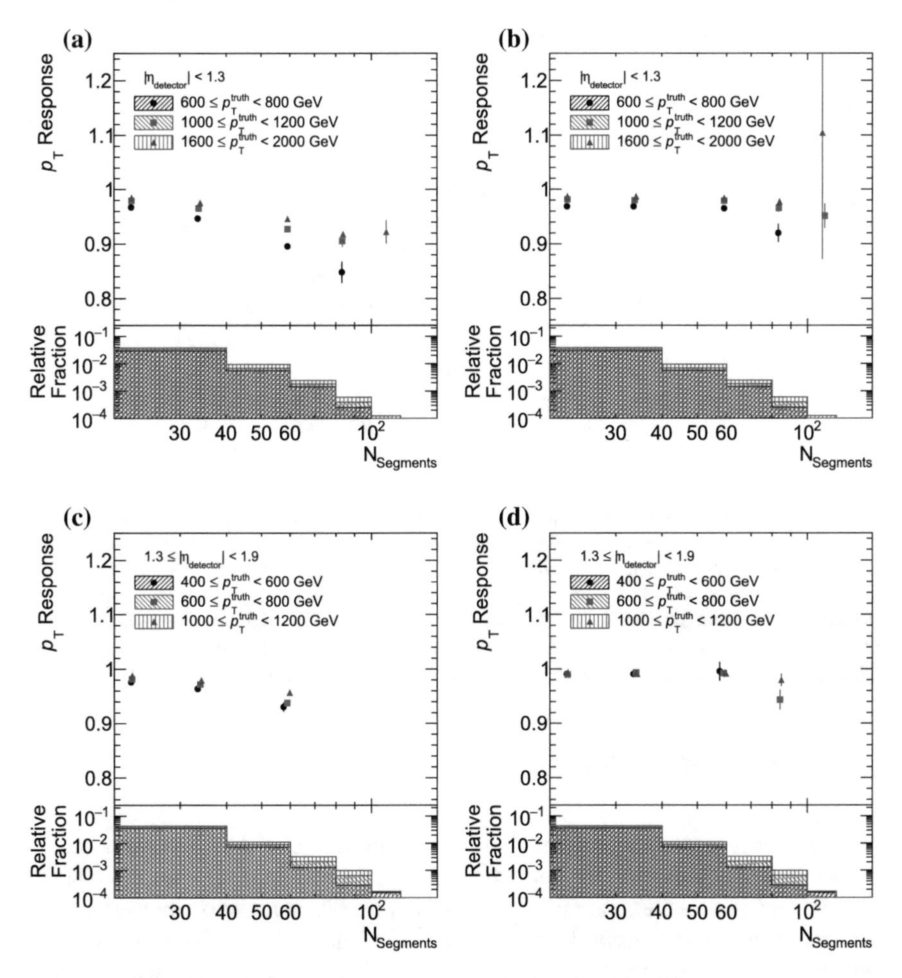

Fig. 4.12 The p_T response as a function of N_{Segments} is shown for three different $p_{T_{\text{truth}}}$ bins (shown in black, red and blue), before the application of the punch-through correction (left), and after the application of the punch-through correction (right), for $|\eta_{\text{detector}}| < 1.3$ (top) and $1.3 \leq |\eta_{\text{detector}}| < 1.9$ (bottom). These figures are adapted from [22]

of additional muon segments from jet punch-through. The plateauing of the b-jet fraction above 20 N_{Segments} indicates that in this region the muon segments from jet punch-through dominate over any additional segments from the semi-leptonic decay of B-hadrons. For this reason, it was decided to utilise a threshold of 20 muon segments, such that only jets with at least 20 N_{Segments} receive the jet punch-through correction.

In the future, a separate punch-through correction could be derived for b-jets and non-b-jets, removing the need for a N_{Segments} threshold and ensuring each type of jet is corrected appropriately for the effects of jet punch-through.

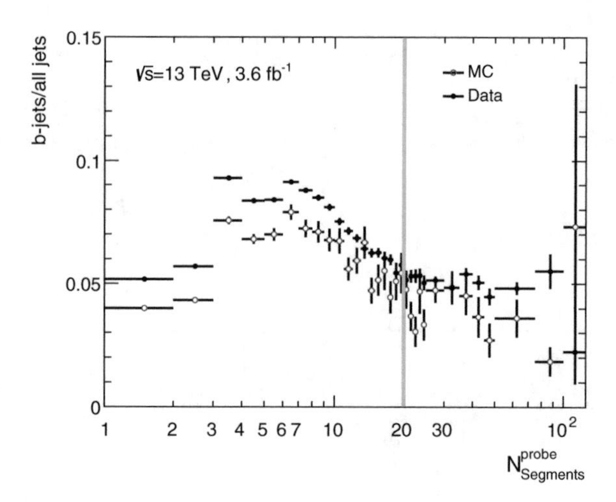

Fig. 4.13 The ratio of b-jets to all jets is shown as a function of $N^{probe}_{Segments}$, for dijet events in which one jet has ≥1 associated muon segment (the probe jet) and the other jet has 0 associated muon segments (the reference jet). There is shown to be a higher fraction of b-jets below ~20 muon segments, with a peak at 3 muon segments, indicating that in this region the increase is due to muon segments from the semi-leptonic decay of B-hadrons. Above ~20 muon segments the ratio plateaus, indicating that in this region the muon segments from jet punch-through dominate. Hence, the jet punch-through correction is applied to jets with $N_{\text{Segments}} \geq 20$

4.2.5 Residual In Situ Calibration

The final stage in the calibration chain is the *residual* in situ *calibration*. Due to the difficulty in modelling the formation of jets, their interaction with the ATLAS detector, and in modelling the detector itself, residual differences in response between MC and data can be present. In data and MC the balance in transverse momentum between jets and other well calibrated objects is used to calculate the average in situ p_T response $R = \langle \frac{p_T^{\text{probe}}}{p_T^{\text{reference}}} \rangle$, where $p_{T^{\text{reference}}}$ is the transverse momentum of the well calibrated object, and p_T^{probe} is the transverse momentum of the jet. The ratio of the response obtained in data to the response obtained in MC $\frac{R_{\text{Data}}}{R_{\text{MC}}}$ is derived as a function of p_T^{probe}, and as a function of $\eta^{\text{probe}}_{\text{detector}}$ for the η-intercalibration. The inverse of this response function provides a correction for the data.

In the residual in situ calibration, several different reference objects are utilised in order to target different detector regions and to span different regions in jet p_T. The correction is derived sequentially, with each correction being applied before the next reference object is utilised. A summary of the reference objects is given below, in the order of application.

η-intercalibration: Dijet events are utilised, in which well calibrated central jets ($|\eta_{\text{detector}}| < 0.8$), are balanced against forward jets ($0.8 < |\eta_{\text{detector}}| < 4.5$). This correction aims to ensure a uniform response to jets in different regions of the detector.

Z/γ-jet balance: In the central region $|\eta_{\text{detector}}| < 0.8$, photons and Z bosons (decaying to e^+e^- or $\mu^+\mu^-$) are used as the reference object to calibrate jets with p_T up to 950 GeV.

Multi-jet balance: Several low-p_T jets (with the full calibration up to the Z/γ-jet balance stage applied) are used as the reference for calibrating jets with p_T up to 2 TeV. The $|\eta_{\text{detector}}| < 1.2$ region is used to derive the calibration.

Note that even though central regions of the detector were used to derive the Z/γ-jet balance and the multi-jet balance corrections, the corrections are also applied in the forward region, since the η-intercalibration equalised the response in both regions.

Since the Z-jet balance, γ-jet balance and multi-jet balance techniques overlap in jet p_T, a combined correction is derived, for full details see [12]. A narrow p_T binning is defined, and in each bin a weighted average of the ratio $\frac{R_{\text{Data}}}{R_{\text{MC}}}$ from each method is calculated. The p_T-dependent weights take into account the relative uncertainty on each result, the original p_T binning used to derive the result, and correlations between p_T bins (systematic uncertainties are treated as fully correlated across p_T and η). The combined result is then lightly smoothed to limit the impact of statistical fluctuations; the inverse of this result is the correction applied to data. Systematic uncertainties are propagated through the procedure and are inflated in regions in which there is tension between the results of the different methods.

The combined result ($\frac{R_{\text{Data}}}{R_{\text{MC}}}$) is shown in Fig. 4.14a for the calibration utilised in the high mass dijet analysis, i.e. the 2012 in situ calibration derived using $\sqrt{s} = 8$ TeV data, and in Fig. 4.14b for the calibration utilised in the dijet + ISR analyses, derived using 2015 $\sqrt{s} = 13$ TeV data. Note that the 2012 in situ ratio shown in Fig. 4.14a was used in combination with additional factors derived using MC, which

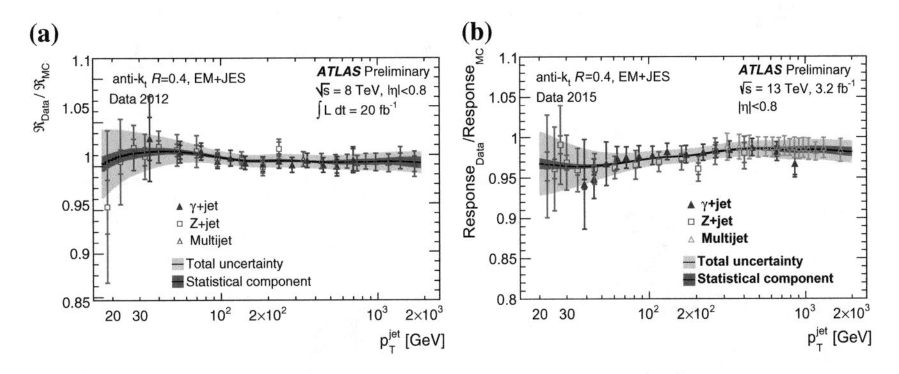

Fig. 4.14 The ratio of the p_T response in data and MC is shown for three of the relative in situ calibrations (γ-jet balance in purple, Z-jet balance in red and multi-jet balance in grey). The final combined result is shown by the black line, and the total uncertainty and statistical component are shown by the green and blue bands, respectively. Figure **a** shows the 2012 in situ calibration [13], utilised in the high mass dijet analysis. Figure **b** shows the 2015 in situ calibration [25], utilised in the dijet + ISR analyses

are not shown in this figure. Figure 4.14 shows that good agreement is observed between each of in situ analyses in both cases. Only one small region of tension was identified for each; for the 2012 in situ calibration the region is at ∼200 GeV, and for the 2015 in situ calibration the region is at ∼50 GeV. The difference in the final result of the combination between the 2012 in situ calibration and the 2015 in situ calibration is expected, based on the simulation changes between Run I and Run II, see [25] for further details.

4.2.6 Jet Energy Scale Uncertainty

Detailed descriptions of the jet energy scale uncertainties are provided in [11] and recommendations from the JetEtmiss performance group are provided in [14, 26] for the high mass dijet analysis, and in [27] for the dijet + ISR analyses. A brief summary of the uncertainties is provided here.

The jet energy scale uncertainty consists of more than 70 nuisance parameters. The majority of these uncertainties are associated with the residual in situ correction; taking into account MC modelling uncertainties, the statistical uncertainty on the samples used to derive the correction, and the uncertainties from the reference objects balanced against the jet. Additional uncertainties are from the flavour composition (fraction of gluons and light quarks in the sample), the response to these objects, pile-up uncertainties, and the jet punch-through uncertainty. Full details about the jet punch-through uncertainty will be given in Sect. 4.3.

Figure 4.15 shows the jet energy scale uncertainty as a function of jet p_T, for (a) the calibration utilised in the high mass dijet analysis, and (b) the calibration utilised in the dijet + ISR analyses. Note that *Absolute* in situ *JES* refers to the Z/γ-jet balance and multi-jet balance calibrations, and *Relative* in situ *JES* refers to the η-intercalibration. Since the in situ calibration applied in the high mass dijet

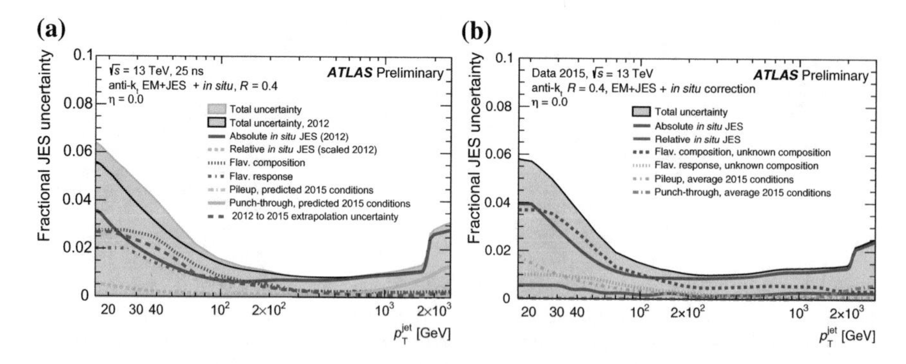

Fig. 4.15 The jet energy scale uncertainty is shown as a function of jet p_T, for central jets with $\eta = 0$, for **a** the calibration utilised in the high mass dijet analysis, and **b** the calibration utilised in the dijet + ISR analyses. Figure **a** is from [14], and Figure **b** is from [25]

search utilised the Run I correction in combination with factors derived in MC there is an additional uncertainty, referred to as the *2012–2015 extrapolation uncertainty*, applied. This uncertainty accounts for the impact of Run I to Run II changes on data and MC.

The total uncertainty ranges from ~6% for 20 GeV jets, down to ~1% for 200 GeV jets, and up to ~3% for 2.5 TeV jets, with a slightly smaller uncertainty for the calibration utilised in the dijet + ISR analyses, particularly in the very high and low mass regions. A sharp rise in the uncertainty is observed at ~1.7 TeV in Fig. 4.15a and at ~2 TeV in Fig. 4.15b. This rise indicates the end point of the multi-jet balance calibration. Above this point, the uncertainty is calculated by propagating single-particle response uncertainties to jets, where the single particle response uncertainties are obtained from Monte Carlo simulations, test-beam studies and data collected by ATLAS [12].

The propagation of more than 70 nuisance parameters in an analysis would be extremely time consuming, and in many cases the loss of correlation information caused by combining nuisance parameters has a negligible effect on the results of the analysis. For this reason, the JetEtmiss performance group combines the nuisance parameters to produce four *strongly-reduced* sets, with each set containing only four nuisance parameters. One of the nuisance parameters is the η-intercalibration non-closure uncertainty, and the other three are combinations of the remaining nuisance parameters. Note that the η-intercalibration non-closure uncertainty was not part of the recommendation for the calibration utilised in the high mass dijet analysis. Hence, each set consisted of only three nuisance parameters and not four.

Each of the sets preserve correlation information in a different area of jet p_T and η phase space. In order to utilise one of the strongly reduced sets instead of the full set of nuisance parameters, analyses must demonstrate that they are insensitive to the loss of correlation information caused by using a strongly-reduced set. Note that the overall size of the JES uncertainty is maintained in the strongly-reduced sets; only the correlation information is different. For details about the construction of the strongly-reduced sets see [28].

4.2.7 Jet Energy Resolution and Uncertainty

As previously mentioned, the jet energy resolution quantifies the spread of the measured jet energies with respect to their true value, indicating how precisely the jet energy can be measured. The determination of the JER is a multi-stage process, which is described in full in [13]. Many of the stages involve using the p_T balance of objects (in dijet events and Z/γ-jet events) to obtain p_T response distributions, in a similar manner to the residual in situ calibration described above. Hence, in practice, rather than calculating $\frac{\sigma(E)}{E}$, we calculate $\frac{\sigma(p_T)}{p_T}$ (still referred to as the JER). The JER is then determined by applying a Gaussian fit to the obtained p_T response distributions, and taking the ratio between the root mean square of the fit $\sigma(p_T)$ and the mean value $\langle p_T \rangle$. The obtained JER as a function of jet p_T is shown in Fig. 4.16a.

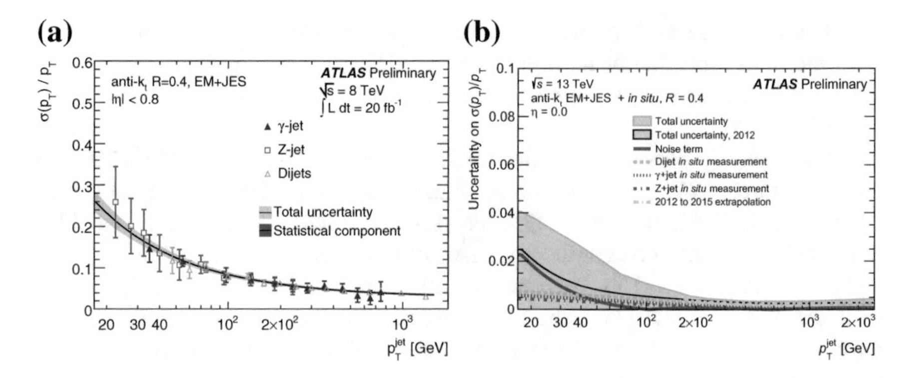

Fig. 4.16 This figure shows **a** the jet energy resolution as a function of jet p_T, and **b** the associated uncertainty as a function of jet p_T. The jet energy resolution is shown for central jets with $|\eta| < 0.8$, and the associated uncertainty is shown for central jets with $\eta = 0$. Figure **a** is from [13], and Figure **b** is from [14]

The uncertainty on the jet energy resolution as a function of jet p_T is shown in Fig. 4.16b. This uncertainty was derived at the time of the high mass dijet analysis using a combination of the Run I uncertainty, and an uncertainty derived by investigating the impact of Run I to Run II changes using Monte Carlo, referred to as the *2012–2015 extrapolation uncertainty*. The same JER uncertainty was utilised in the dijet + ISR analyses as the JER uncertainty derived using Run II data was not yet available. The uncertainty is largest at low p_T due to the uncertainties on noise contributions and the extrapolation uncertainty. Since the extrapolation uncertainty is dominant, the JER nuisance parameters were combined to form a single nuisance parameter [14].

4.3 Jet Punch-Through Uncertainty

In Sect. 4.2.6, the uncertainties associated with the jet energy scale correction were shown. An important uncertainty for high p_T jets is the uncertainty associated with the jet punch-through correction. Since the jet punch-through correction is derived using Monte Carlo, an uncertainty is needed to cover the differences between data and Monte Carlo. In this section, the derivation of the jet punch-through uncertainty is described. This uncertainty was derived for anti-k_t jets with $R = 0.4$ using the full 2015 $\sqrt{s} = 13$ TeV data set (3.6 fb^{-1}), and formed part of the 2016 jet energy scale uncertainties, utilised by all ATLAS analyses which include jets. This includes the dijet + ISR analyses described in this thesis.

Since the purpose of the uncertainty is to cover differences between data and Monte Carlo, when deriving the uncertainty we must use in situ quantities, i.e. only information which is available in data, rather than truth level quantities, for example.

A tag-and-probe method using dijet events was chosen, exploiting the p_T balance between the two jets. Note that such balance techniques are utilised in other stages of the JES calibration and its corresponding uncertainties. Events with a dijet topology were selected using the criteria given in Table 4.1. This selection is similar to the high mass dijet analysis selection, which is given in Table 5.2; see this chapter for further details about each requirement. Some additional selections are applied here in order to ensure that we are selecting well balanced dijet events. The requirement on $|\Delta\phi|$ is to ensure that the jets are back-to-back, rejecting multi-jet like events. The requirement on the third jet p_T also helps to reject multi-jet events in which the third jet p_T is relatively high. The leading jets are required to be within $|\eta_{\text{detector}}| < 2.7$ as this is the acceptance region of the muon spectrometer. In the derivation of the uncertainty the jets are calibrated up to the global sequential calibration level, including the jet punch-through correction, but no insitu correction is applied (the exception being in the application of the analysis selection, in which the insitu calibration is applied).

For events with exactly two perfectly calibrated jets, we expect to see a complete balance in p_T between the two jets that form the dijet. This scenario is illustrated in Fig. 4.17. The impact of the punch-through correction can be assessed by using events where one of the jets in the dijet is a punch-through jet ($N_{Segments} \geq 1$), and the other jet does not punch-through ($N_{Segments} = 0$). A tag-and-probe technique is then used, where the punch-through jet is chosen as the probe jet, and the jet which does not punch-through is the tag (reference) jet. The difference between the average in situ p_T response $\langle \frac{p_T^{probe}}{p_T^{reference}} \rangle$ in data and in Monte Carlo provides us with the uncertainty.

The end goal is for users to be able to pass a punch-through corrected jet to the `JetUncertaintyProvider`, and to obtain the uncertainty due to the jet punch-

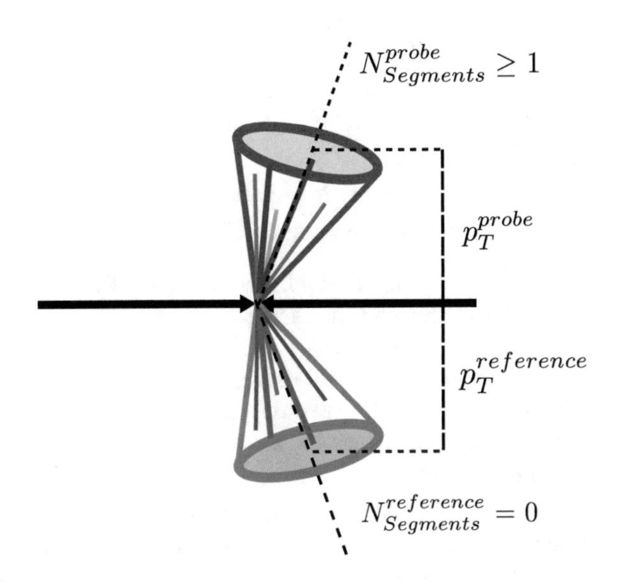

Fig. 4.17 Schematic illustration of the two jets in a dijet system balancing in p_T, with the jet with $N_{Segments} = 0$ being the tag (reference) jet, and the jet with $N_{Segments} \geq 1$ being the probe jet. Figure adapted from [9, 17]

through correction for this jet. The uncertainty provided will depend on the properties of the jet, i.e. the jet p_T, η_{detector} and N_{Segments}. The uncertainty must be derived across the full phase space, i.e. all p_T, η_{detector} and N_{Segments} values which jets could have.

An overview of the steps involved in deriving the uncertainty will now be given, before providing more details about each step.

1. The phase space is divided into bins of $\eta^{probe}_{\text{detector}}$, $N^{probe}_{Segments}$, and $p_T^{average}$, where $p_T^{average} = \frac{1}{2}(p_T^{probe} + p_T^{reference})$. The following bins were utilised:
   η_{detector}: 0, 1.3, 1.9, 2.7
   $N_{Segments}$: 0, 1, 5, 10, 15, 20, 25, 50, 75, 100, 150, 200
   $p_T^{average}$: 420, 600, 1000, 1600, 3000, 4500

2. For each bin, the following quantities were calculated:
   - The mean of the $N^{probe}_{segments}$ distribution (calculated using the ROOT `GetMean` function).
   - The mean of the p_T^{probe} distribution (calculated using the ROOT `GetMean` function).
   - The mean of the asymmetry distribution, where the asymmetry A is given by $A = \frac{p_T^{probe} - p_T^{reference}}{p_T^{average}}$ (obtained via fitting a Gaussian function to the distribution).

3. In situ p_T response $(\langle \frac{p_T^{probe}}{p_T^{reference}} \rangle = \frac{2+\langle A \rangle}{2-\langle A \rangle})$ versus p_T^{probe} versus $N^{probe}_{segments}$ 2D graphs were constructed in bins of $\eta^{probe}_{\text{detector}}$, for data and MC.
4. The 2D graph was converted into a 2D histogram using the ROOT `Interpolate` function.
5. The uncertainty is given by the data-MC difference:
   $\left| \left\langle \frac{p_T^{probe}}{p_T^{reference}} \right\rangle_{Data} - \left\langle \frac{p_T^{probe}}{p_T^{reference}} \right\rangle_{MC} \right|$.
6. In order to cover the full phase space, the uncertainty is extrapolated. First we extrapolate to higher p_T^{probe} values, then to lower p_T^{probe} values, then to higher $N^{probe}_{segments}$ values.
7. The final uncertainty is provided as a 3D histogram with p_T^{probe} on the x-axis, $N^{probe}_{segments}$ on the y-axis, and $\eta^{probe}_{\text{detector}}$ z-axis.

The selected binning in step 1 is based on the Run I binning described in [17]. Each η_{detector} bin corresponds to a separate region of the detector, i.e. $|\eta_{\text{detector}}| < 1.3$ is the barrel region, $1.3 \le |\eta_{\text{detector}}| < 1.9$ spans the poorly instrumented transition region between the barrel and end-cap, and $1.9 \le |\eta_{\text{detector}}| < 2.7$ is the hadronic end-cap region. From Fig. 4.10a, we observed that the level of data-MC agreement is different in each of these three regions, with worse agreement being observed for the two more forward η_{detector} bins. The $N^{probe}_{segments}$ and $p_T^{average}$ binning is finer at low $N^{probe}_{segments}$ and low $p_T^{average}$, respectively, increasing in the higher $N^{probe}_{segments}$ and $p_T^{average}$ regions, respectively, where there are limited numbers of events. The $p_T^{average}$ starting point at 420 GeV was selected as it is above the leading jet p_T cut of 410 GeV. It was decided to bin in $p_T^{average}$, rather than p_T^{probe} or $p_T^{reference}$ as this reduces the impact of resolution differences between the probe and the reference jet.

In step 2, we obtain the mean of the $N_{segments}^{probe}$ distribution, the p_T^{probe} distribution, and the $A+1$ distribution. These quantities are needed in order to calculate the response (using $\langle A \rangle$) as a function of $N_{segments}^{probe}$ and p_T^{probe} in the next step. By obtaining the mean of the p_T^{probe} distribution, this provides us with a mapping between the $p_T^{average}$ bin and the average p_T^{probe} in this bin. This is important as the uncertainty must be parametrised in terms of p_T^{probe}, as this is the jet property which will be passed to the `JetUncertaintyProvider`. As previously mentioned, the mean of the $A+1$ distribution $\langle A+1 \rangle$ is obtained via fitting a Gaussian function to the distribution. The fit is performed using the `JES_ResponseFitter` package [29], which is a package developed by the JetEtmiss performance group for the fitting of response distributions. The quantity $A+1$ is utilised, rather than just A, as the fitting package expects a distribution centered around one, not zero. The fit is performed over the core of the distribution ($\langle A+1 \rangle \pm 1.6\sigma$ from the mean). Examples of $A+1$ distributions and their corresponding fits are shown in Fig. 4.18. For further details about the fitting procedure see Sect. 5.4 of [17]. By utilising a fit to the core of the distribution, we obtain the value of the mean without taking into account any tails which may be present in the distribution. Once $\langle A+1 \rangle$ is obtained, we subtract 1 to

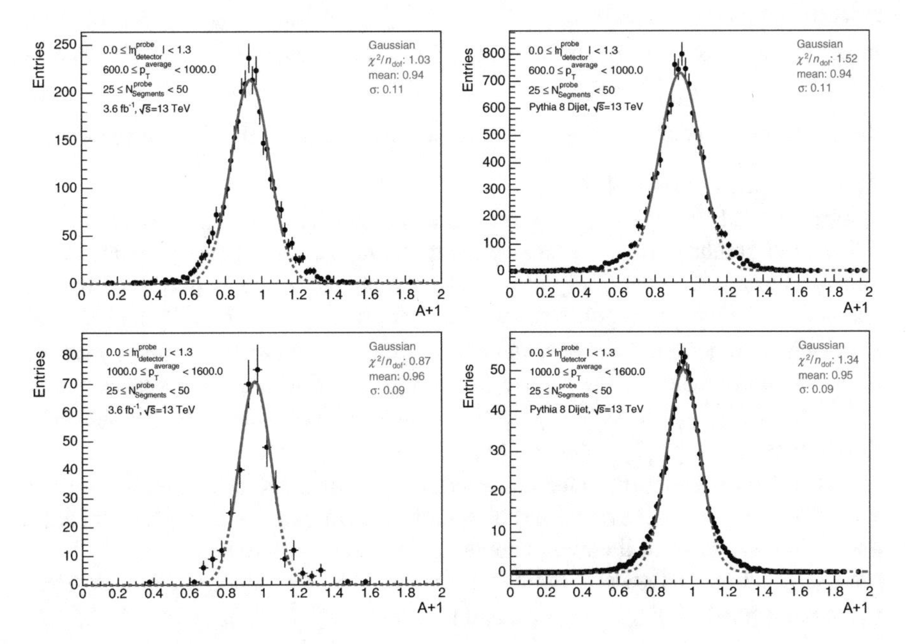

Fig. 4.18 Examples of fits to the asymmetry + 1 distribution (A + 1) in data (left) and Monte Carlo (right), in bins of $|\eta_{\text{detector}}|$, $N_{\text{Segments}}^{\text{probe}}$ and p_T^{average}, for two different p_T^{average} bins (top and bottom). The solid line shows the fit in the region $\langle A+1 \rangle \pm 1.6\sigma$

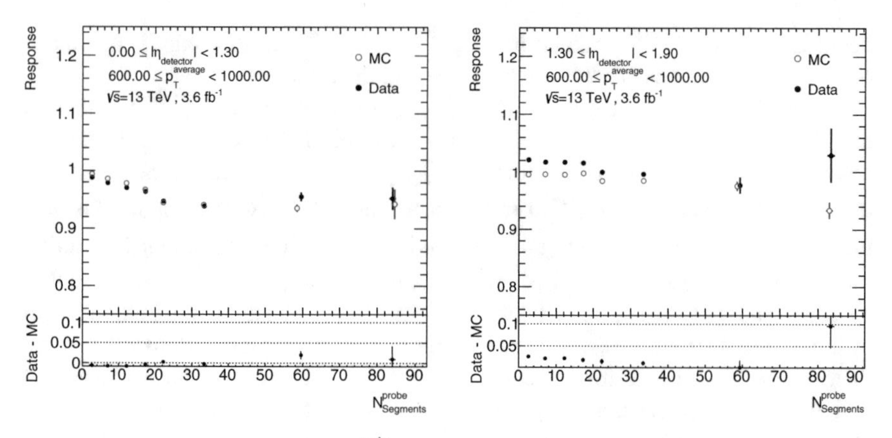

Fig. 4.19 The in situ p_T response ($\frac{p_T^{probe}}{p_T^{reference}}$) is shown as a function of $N_{segments}^{probe}$ for two different η_{detector} bins (and the same p_T^{average} bin)

obtain the mean asymmetry $\langle A \rangle$, which will be used to calculate the response in the next step.

In step 3, the mean values obtained in the previous step are used to construct 2D response graphs as a function of p_T^{probe} and $N_{segments}^{probe}$, in bins of $\eta_{\text{detector}}^{probe}$. A projection of the 2D response graphs in the plane of response versus $N_{segments}^{probe}$ is shown in Fig. 4.19 for two $\eta_{\text{detector}}^{probe}$ bins, as an example. The observed trends shown here for the in situ p_T response ($\frac{p_T^{probe}}{p_T^{reference}}$) reflect those seen earlier in the post-correction p_T response ($\frac{p_T^{reco}}{p_T^{truth}}$), in Fig. 4.12b, d.

For the final uncertainty, we want more detailed information, so in step 4 we convert our graphs into histograms by interpolating between the graph points. This allows us to choose how finely we wish to sample the values between the points, enabling us to derive a response curve with finer granularity. The p_T^{probe} axis is sampled with increments of 100 GeV, and the $N_{segments}^{probe}$ axis is sampled with increments of 10 muon segments. We now have 2D histograms of in situ p_T response versus p_T^{probe} and $N_{segments}^{probe}$ for each $\eta_{\text{detector}}^{probe}$ bin, in data and MC. An example is shown in Fig. 4.20 for the $0 \leq |\eta_{\text{detector}}^{probe}| < 1.3$ bin.

In step 5, we calculate the uncertainty by taking the data-MC difference in response for regions where we have both data and Monte Carlo response values. The calculated uncertainty is shown in the region contained within the white box in Fig. 4.21, for each $\eta_{\text{detector}}^{probe}$ bin. The general trends observed in the uncertainty is that it is larger in the more forward $\eta_{\text{detector}}^{probe}$ regions, and for higher $N_{Segments}^{probe}$, as expected, and it is larger for higher p_T^{probe}. The largest uncertainty observed in each $\eta_{\text{detector}}^{probe}$ bin is as follows: 4.7% for the most central $\eta_{\text{detector}}^{probe}$ bin, 8.7% for the more forward eta bin, and 9.3% for the most forward $\eta_{\text{detector}}^{probe}$ bin.

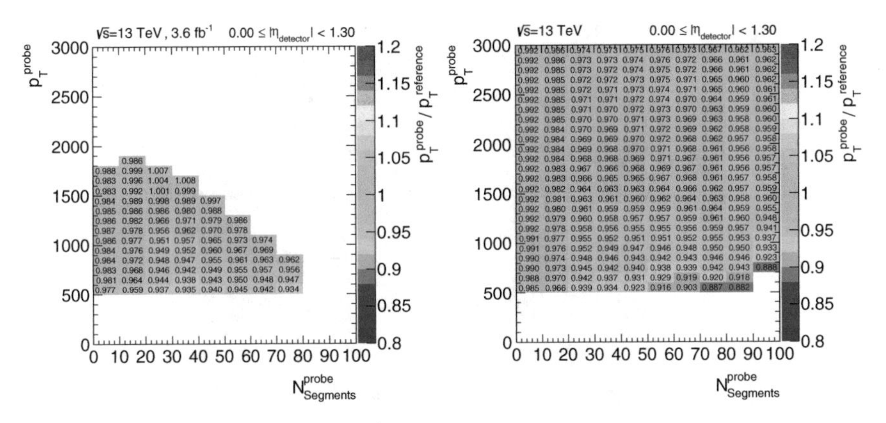

Fig. 4.20 The in situ p_T response ($\frac{p_T^{probe}}{p_T^{reference}}$) is shown in the plane of p_T^{probe} versus $N_{Segments}^{probe}$ for the central region $0 \leq |\eta_{\text{detector}}^{probe}| < 1.3$, for data (left) and Monte Carlo (right)

As previously described, the uncertainty must span the full phase space, so in step 6 we extrapolate the derived uncertainty. A simple extrapolation was used, in which the calculated uncertainty values at the edge of the white box were used in other regions of phase space. The values in the top rows were first extended to higher values of p_T^{probe}, then the values in the lowest rows were extended to lower values of p_T^{probe}, and finally the columns furthest to the right were extended to higher values of $N_{segments}^{probe}$. The extrapolated uncertainties are shown in Fig. 4.21.

In step 7, we obtain the final uncertainty in a 3D histogram, by effectively stacking up the 2D uncertainty histograms from each $\eta_{\text{detector}}^{probe}$ bin. The final uncertainty provided spans the full phase space, i.e. $N_{segments}^{probe}$ up to 250, p_T^{probe} up to 7 TeV and all $\eta_{\text{detector}}^{probe}$ bins. In the future, a more precise extrapolation procedure could be utilised, such as the nonparametric regression technique utilised in Run I [17]. Additionally, the extrapolation could be performed across the $\eta_{\text{detector}}^{probe}$ bins.

The method used to derive the uncertainty in Run II differs to the method used in Run I, which is documented in detail in [17]. In Run I the uncertainty was derived in bins of p_T^{probe}, rather than $p_T^{average}$, and hence, the mapping from $p_T^{average}$ to p_T^{probe} was not needed and the uncertainty was derived in one dimension as a function of $N_{segments}^{probe}$. In addition, the response $\frac{p_T^{probe}}{p_T^{reference}}$ was calculated and fitted directly, rather than being calculated via the asymmetry. In Run II, prior to these changes, when binning in p_T^{probe} rather than $p_T^{average}$, tails were observed in the response distribution, with an excess at high response values. The changes were made following advice from members of the η-intercalibration team, and Gaussian shaped asymmetry distributions with symmetric tails were obtained when using the recommended procedure.

As previously mentioned, when deriving the punch-through correction, there were insufficient events above $\eta_{\text{detector}} = 1.9$; thus, there is no jet punch-through correction

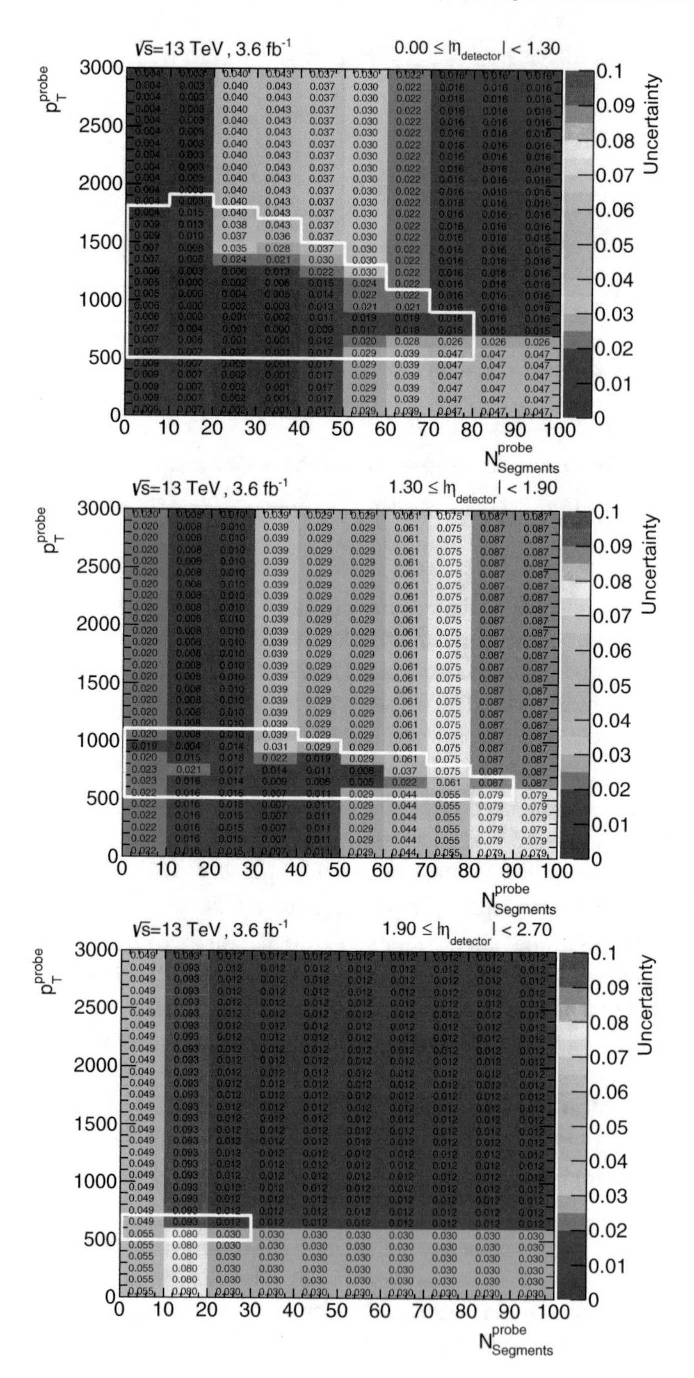

Fig. 4.21 The white box shows the region where the jet punch-through uncertainty is derived. These uncertainty values are extrapolated to the regions outside the white box. The uncertainty values are shown in the plane of p_T^{probe} versus $N_{Segments}^{probe}$, in bins of $\eta_{\text{detector}}^{probe}$

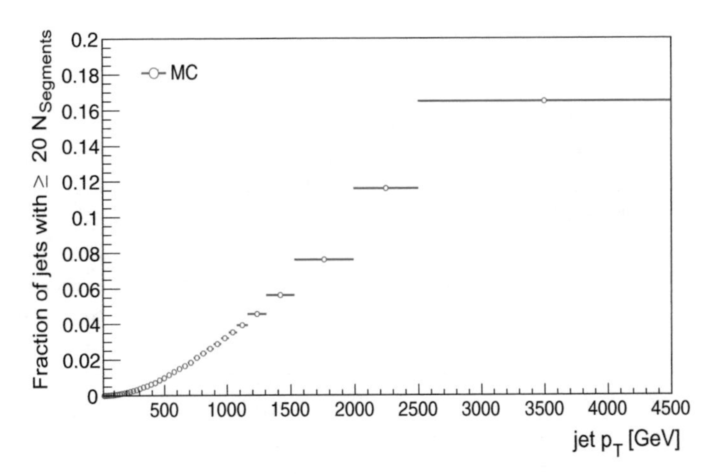

Fig. 4.22 The fraction of jets with ≥ 20 N_{Segments}, i.e. the fraction receiving the jet punch-through correction as a function of jet p_T

applied in this region. It was decided to apply the derived uncertainty in this region, even though there is no correction, as the uncertainty provides a measure of the impact of jet punch through in this region. In contrast, it was decided not to apply the derived uncertainty to jets with $N^{probe}_{segments} < 20$, which do not receive the jet punch-through correction. The uncertainty is not applied in order to avoid applying an unnecessary extra uncertainty to jets containing B-Hadrons.

In Fig. 4.15, the jet punch-through uncertainty was shown for jets with $\eta = 0$, for the calibration utilised in the high mass dijet analysis, and for the calibration utilised in the dijet + ISR analyses. It should be noted that when producing this plot, the derived uncertainty is multiplied by the average fraction of jets receiving the punch-through correction (i.e. those with $N_{\text{Segments}} \geq 20$), to give a realistic estimate of the impact of this uncertainty. This fraction is shown in Fig. 4.22 as a function of jet p_T.

By comparing Fig. 4.15a, b, we see that the jet punch-through uncertainty is larger for the calibration utilised in the high mass dijet analysis, than for the calibration utilised in the dijet + ISR analyses. For the calibration utilised in the high mass dijet analysis, there was insufficient Run II data to derive the uncertainty using data. The initial plan was to utilise the Run I uncertainty plus additional factors to cover the changes between Run I and Run II. However, due to the change in N_{Segments} observed between Run I and Run II, shown previously in Fig. 4.11a, the use of the Run I uncertainty would require a 'mapping' between N_{Segments} in Run I and N_{Segments} in Run II. This approach was dropped, as the relationship is not expected to be one-to-one, making the mapping extremely complex. Instead, a conservative approach was taken in which the maximum Run I uncertainty (10%) [17] was taken as a flat uncertainty for all jets with $N_{\text{Segments}} \geq 20$. For the calibration utilised in the dijet + ISR analyses, the uncertainty derived in this section was utilised. Hence, we observe a reduction in the uncertainty with respect to the uncertainty utilised in the high mass dijet analysis.

4.4 Photon Reconstruction

As previously described, upon interaction with the calorimeter, highly energetic photons produce a shower of lower energy electrons, positrons and photons, depositing a large amount of energy in the EM calorimeter. In order to reconstruct photons, a combination of calorimeter information and tracking information is used. An overview of photon reconstruction will be provided here, for full details see [4, 30, 31].

The first step in the photon reconstruction is to identify groups of EM calorimeter cells which contain a significant energy deposit, these regions are referred to as *seed clusters*. In order to do this, within $|\eta| < 2.47$, the EM calorimeter is divided into a grid of 200×256 calorimeter *towers*, with a size of $\Delta\eta \times \Delta\phi = 0.025 \times 0.025$ in η and ϕ (matching the granularity of the EM calorimeter cells in the middle layer). The sum of the energy contained in each tower is computed, and a fixed-size sliding window technique [4] is applied to locate local maxima in transverse energy E_T, which exceed 2.5 GeV. The sliding window technique involves scanning over the tower grid in windows of 3×5 towers. Once local maxima have been located, a more precise position calculation is performed and seed clusters are defined.

The next step involves identifying the seed clusters as either electrons, converted photons or unconverted photons, where converted means that the photon has transformed into an electron-positron pair [31]. In order to perform this categorisation, a loose matching of inner detector tracks to seed clusters is performed using geometrical requirements. The tracks are then used to identify single or double track photon conversion vertices. For double track conversion vertices, the tracks must be consistent with opposite sign electrons, based on hit requirements in the TRT. For single track conversion vertices, the track must be consistent with an electron and have no hits in the IBL. The conversion vertices are then matched to the cluster using geometrical requirements. If a cluster has no matched track or conversion vertex then it is classified as an unconverted photon. If a cluster is matched to either a single or double track conversion vertex then it is classified as a converted photon.

Once classified, the final calorimeter clusters are built with a size that depends on the position of the cluster. For converted and unconverted photons in the EM barrel a 3×7 cluster is built (measured in units of towers), and in the EM end-cap a cluster size of 5×5 is utilised [30]. The size of the clusters is a compromise between minimising noise and containing the full energy of the photon.

The energy calculated for the final clusters is then calibrated in order to correct for effects such as energy losses in material before reaching the calorimeter, or energy leaking outside the cluster [30]. A separate calibration is derived for converted photons and unconverted photons, due to their differences in material interactions in the inner detector [32]. The latest calibration provided by the ATLAS EGamma performance group was used; for full details about the calibration and associated uncertainty see [30, 33].

References

1. Blazey GC et al (2000) Run II jet physics. In: QCD and weak boson physics in Run II. Proceedings, Batavia, USA, 4–6 March, 3–4 June, 4–6 November 1999. pp 47–77. arXiv: hep-ex/0005012 [hep-ex]
2. Kirschenmann H (2012) Jets at CMS and the determination of their energy scale. http://cms.web.cern.ch/news/jets-cms-and-determination-their-energy-scale
3. ATLAS Collaboration (2016) Topological cell clustering in the ATLAS calorimeters and its performance in LHC Run 1. arXiv: 1603.02934 [hep-ex]
4. Lampl W et al (2008) Calorimeter clustering algorithms: description and performance. Technical report, ATL-LARG-PUB-2008-002. Geneva: CERN
5. Boelaert N (2012) Dijet angular distributions in proton-proton collisions at $\sqrt{s} = 7$ TeV and $\sqrt{s} = 14$ TeV. Ph.D. thesis. https://doi.org/10.1007/978-3-642-24597-8, ISBN: 9783642245978
6. Cacciari M, Salam GP, Soyez G (2008) The anti-kt jet clustering algorithm. J High Energy Phys 2008(04):063. https://doi.org/10.1088/1126-6708/2008/04/063
7. Rojo J (2014) Lecture notes from 'The strong interaction and LHC phenomenology' course. Oxford. http://www2.physics.ox.ac.uk/sites/default/files/2014-03-31/qcdcourse_juanrojo_tt2014_lect8_pdf_18445.pdf
8. Sapeta S (2016) QCD and jets at hadron colliders. Prog Part Nucl Phys 89:1–55. https://doi.org/10.1016/j.ppnp.2016.02.002, arXiv: 1511.09336 [hep-ph]
9. Doglioni C (2012) Measurement of the inclusive jet cross section with the ATLAS detector at the large hadron collider. Springer Theses. Springer, Berlin. https://doi.org/10.1007/978-3-642-30538-2, ISBN: 9783642305382
10. Cacciari M, Salam GP (2008) Pileup subtraction using jet areas. Phys Lett B659:119–126. https://doi.org/10.1016/j.physletb.2007.09.077, arXiv: 0707.1378 [hep-ph]
11. ATLAS Collaboration (2017) Jet energy scale measurements and their systematic uncertainties in proton-proton collisions at $\sqrt{s} = 13$ TeV with the ATLAS detector. Phys Rev D96:072002. https://doi.org/10.1103/PhysRevD.96.072002, arXiv: 1703.09665 [hep-ex]
12. ATLAS Collaboration (2013) Jet energy measurement with the ATLAS detector in proton-proton collisions at $\sqrt{s} = 7$ TeV. Eur Phys J C73.3:2304. https://doi.org/10.1140/epjc/s10052-013-2304-2, arXiv: 1112.6426 [hep-ex]
13. ATLAS Collaboration (2015) Monte Carlo calibration and combination of in-situ measurements of jet energy scale, Jet energy resolution and jet mass in ATLAS. Technical report, ATLAS-CONF-2015-037. Geneva: CERN
14. ATLAS Collaboration (2015) Jet calibration and systematic uncertainties for jets reconstructed in the ATLAS detector at $\sqrt{s} = 13$ TeV. Technical report, ATLAS-PHYS-PUB-2015-015. Geneva: CERN
15. ATLAS Collaboration (2016) Performance of pile-up mitigation techniques for jets in pp collisions at $\sqrt{s} = 8$ TeV using the ATLAS detector. Eur Phys J C76.11:581. https://doi.org/10.1140/epjc/s10052-016-4395-z, arXiv: 1510.03823 [hep-ex]
16. Cacciari M, Salam GP, Soyez G (2008) The catchment area of jets. JHEP 04:005. https://doi.org/10.1088/1126-6708/2008/04/005, arXiv: 0802.1188 [hep-ph]
17. Gupta S (2015) A study of longitudinal hadronic shower leakage and the development of a correction for its associated effects at $\sqrt{s} = 8$ TeV with the ATLAS detector. CERN-THESIS-2015-332. Ph.D. thesis. The University of Oxford (2015)
18. Gupta S, Issever C, Doglioni C (2013) Jet punch-through studies at $\sqrt{s} = 8$ TeV with the ATLAS detector. Technical report, ATL-COM-PHYS-2013-311. Geneva: CERN
19. ATLAS Collaboration, Marshall Z (2014) Simulation of pile-up in the ATLAS experiment. J Phys Conf Ser 513:022024. https://doi.org/10.1088/1742-6596/513/2/022024
20. Fabjan CW, Ludlam T (1982) Calorimetry in high-energy physics. Ann Rev Nucl Part Sci 32:335–389. https://doi.org/10.1146/annurev.ns.32.120182.002003
21. ATLAS Collaboration (2008) The ATLAS experiment at the CERN large hadron collider. J. Instrum. 3.08:S08003. https://doi.org/10.1088/1748-0221/3/08/S08003

22. Bossio J (2017) Private communication
23. ATLAS Collaboration (2016) Performance of b-jet identification in the ATLAS experiment. JINST 11.04:P04008. https://doi.org/10.1088/1748-0221/11/04/P04008, arXiv:1512.01094 [hep-ex]
24. ATLAS Collaboration (2016) Optimisation of the ATLAS b-tagging performance for the 2016 LHC Run. Technical report, ATL-PHYS-PUB-2016-012. Geneva: CERN
25. ATLAS Collaboration (2016) Public plots: Jet energy scale uncertainties updated for ICHEP 2016 using full 13 TeV 2015 dataset. https://atlas.web.cern.ch/Atlas/GROUPS/PHYSICS/PLOTS/JETM-2016-010
26. ATLAS Collaboration (2015) Recommendations for early 2015 analysis with pre-recommendation xAOD calibrations. https://twiki.cern.ch/twiki/bin/viewauth/AtlasProtected/JetUncertainties2015Prerec
27. ATLAS Collaboration (2016) Uncertainty release for analyses using ICHEP2016 calibration version. https://twiki.cern.ch/twiki/bin/viewauth/AtlasProtected/JetUncertainties2015ICHEP2016
28. ATLAS Collaboration (2015) A method for the construction of strongly reduced representations of ATLAS experimental uncertainties and the application thereof to the jet energy scale. Technical report, ATL-PHYS-PUB-2015-014. Geneva: CERN
29. ATLAS JetEtmiss performance group (2017) JES_ResponseFitter. https://svnweb.cern.ch/trac/atlasperf/browser/CombPerf/JetETMiss/JetCalibrationTools/DeriveJES/trunk/JES_ResponseFitter
30. ATLAS Collaboration (2016) Electron and photon energy calibration with the ATLAS detector using data collected in 2015 at $\sqrt{s} = 13$ TeV. Technical report, ATLAS-PHYS-PUB-2016-015. Geneva: CERN
31. ATLAS Collaboration (2016) Measurement of the photon identification efficiencies with the ATLAS detector using LHC Run-1 data. Eur Phys J C 76.12:666. https://doi.org/10.1140/epjc/s10052-016-4507-9, arXiv: 1606.01813 [hep-ex], ISSN: 1434-6052
32. Hance M (2012) Photon physics at the LHC: a measurement of inclusive isolated prompt photon production at $\sqrt{s} = 7$ TeV with the ATLAS detector (Springer Theses). Springer, Berlin. http://www.springer.com/gp/book/9783642330612, ISBN 9783642330629
33. ATLAS Collaboration (2014) Electron and photon energy calibration with the ATLAS detector using LHC Run 1 data. Eur Phys J C 74.10:3071. https://doi.org/10.1140/epjc/s10052-014-3071-4, ISSN: 1434-6052

Chapter 5
Dijet Invariant Mass Spectra

The well calibrated jets can now be used to construct dijet invariant mass spectra. New BSM particles or interactions can then be searched for by looking for resonance 'bumps' above the smoothly falling spectra predicted by QCD. In order to perform such searches experimentally, events which contain at least two jets are selected. The invariant mass of the pair of jets (dijet) which is most likely to have been produced by the decay of the resonance is then calculated by summing together the calibrated four-momentum vectors of the two jets, which is equivalent to the equation

$$m_{jj} = \sqrt{(E_1 + E_2)^2 - (\vec{p}_1 + \vec{p}_2)^2}, \tag{5.1}$$

where E_1 (E_2) is the energy of the leading (sub-leading) jet and $\vec{p}_1$ ($\vec{p}_2$) is the momentum of the leading (sub-leading) jet, if the dijet invariant mass is calculated using the two leading jets. This quantity is calculated for each of the selected events, such that a spectrum of number of events versus dijet invariant mass can be produced.

Expanding Eq. (5.1), and using the relations $E_1 = \sqrt{m_1^2 + p_1^2}$ and $E_2 = \sqrt{m_2^2 + p_2^2}$ yields the equation

$$m_{jj} = \sqrt{m_1^2 + m_2^2 + 2(E_1 E_2 - |\vec{p}_1||\vec{p}_2| cos(\theta))}, \tag{5.2}$$

where θ is the angle between the momenta of the two jets. From this equation, we can see that there is a relationship between the dijet invariant mass m_{jj} and the angle between the two jets. Typically a larger m_{jj} corresponds to a larger separation angle. However, care must be taken as Lorentz boosts to the system can change both the energy and momenta of the jets, as well as the separation angle between them, while m_{jj} is preserved.

L. A. Beresford, *Searches for Dijet Resonances*, Springer Theses,
https://doi.org/10.1007/978-3-319-97520-7_5

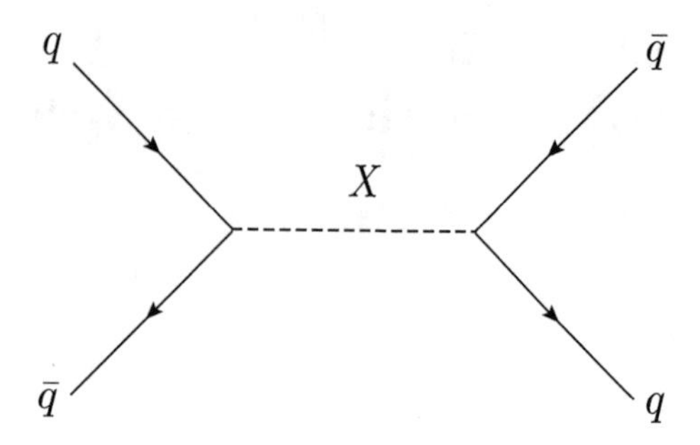

Fig. 5.1 Example Feynman diagram showing the production of a new resonance X decaying to the dijet final state

In order to produce dijet invariant mass spectra for the three analyses described in this thesis, decisions need to be taken about the strategy for including data in the analysis, and which events are suitable for inclusion. The data inclusion strategy, datasets, simulated samples and analysis selection are described in Sect. 5.1 for the high mass dijet analysis, and in Sect. 5.2 for the dijet + ISR analyses. Once the selection is defined data-MC cross-checks are performed in the phase space of the analysis to ensure that there are no unforeseen issues with the data; a small selection of such cross checks are shown in Sect. 5.3. Finally, in order to produce the spectra the binning for each spectrum needs to be chosen. In Sect. 5.4 the binning for the spectra is derived and the final mass spectra are presented.

5.1 The High Mass Dijet Analysis

The high mass dijet analysis was the first of the three analyses to be performed, and it was the first ATLAS search paper to be published using $\sqrt{s} = 13$ TeV data. The paper was published in Physics Letters B in March 2016 [1]. As this analysis aims to search for heavy new resonances the invariant mass is calculated using the two highest p_T jets in the event. At such high masses (1.1 TeV and above) it is extremely unlikely that a jet originating from pile-up or initial state radiation could be produced with higher p_T than the jets from the decay of a resonance. An example diagram showing the production of a new resonance and its decay to the dijet final state is shown in Fig. 5.1.

5.1.1 Blinding Strategy

The data collected by ATLAS in 2015 was the first $\sqrt{s} = 13$ TeV data produced by the LHC, and the high mass dijet analysis was one of the first analyses to analyse the $\sqrt{s} = 13$ TeV data. Prior to analysing this data, a decision needs to be made about the strategy for including data as it is collected. The decision was made to perform this analysis *un-blinded*. The analysis selection and background estimation strategy was fixed prior to data taking, and the analysis was performed for each substantial increase in dataset size, rather than performing the analysis on data only once the full dataset was collected.

Performing the analysis un-blinded is beneficial for several reasons. One key benefit is that this strategy allows any issues with the new data to be identified and solved early on, reducing any loss of data. Due to this approach, the high mass dijet analysis was the first analysis to observe a loss in the number of high mass dijet events. The reason for this loss was due to a problem with the configuration of the trigger algorithm for saturated trigger towers. Trigger towers become saturated when a transverse energy of approximately 250 GeV or higher is deposited [2]. The issue caused the trigger to select the bunch crossing prior to the collision containing the high mass dijet event for the affected events. For the early 2015 data, 50 ns bunch spacing was used, hence, the prior bunch crossing was empty. Fortunately, as the problem was identified and resolved early on, it only resulted in the loss of ~80 pb^{-1} of data.

Another major benefit of the un-blinded strategy is that by analysing larger datasets as soon as they are available, we would be able to make an early discovery if a signal is present. The choice of an un-blinded strategy is also beneficial for the background estimation strategy, as will be described in Chap. 6.

5.1.2 Dataset and Simulated Samples

The high mass dijet analysis utilised the full 2015 dataset, consisting of 3.6 fb^{-1} data. This data includes the events recorded to the debug stream. The fraction of events in the debug stream was seen to increase with dijet invariant mass, highlighting the importance of including the debug stream events corresponding to each data taking run included in the analysis. The increase with dijet mass is thought to be due to the timing out of the muon reconstruction system due to increased jet punch-through. The majority of the data included in the analysis was collected when the full detector was operating well; however, a small amount of data was collected when particular sub-detectors were not operating correctly, or running conditions were altered. Before inclusion of this 'non-standard data', cross-checks were made to ensure that differences between the standard data and the non-standard data are negligible, or covered by the systematic uncertainties. A brief review of the non-standard data will now be given:

- Approximately 25 pb^{-1} of data (less than 1% of the full dataset) was collected using 50 ns spacing between the proton bunches (the standard data utilises 25 ns spacing). Differences in jet p_T and jet η when using QCD dijet MC with 25 ns bunch spacing, and QCD dijet MC with 50 ns bunch spacing were shown to be negligible in the phase space of the analysis. The difference in the jet energy scale uncertainty for 25 ns data and for 50 ns data is less than 1%, and is neglected.

- Approximately 280 pb^{-1} of data ($\sim$8% of the full dataset) was collected when the Insertable B-Layer (IBL) was off, or in standby mode. The data collected with the IBL on was found to be consistent with the data collected with the IBL off, within statistical uncertainties, both with and without the global sequential calibration applied (this calibration utilises tracking information).
- For 120 events in the analysis, an issue with one of the trigger algorithms caused the trigger to select the bunch crossing prior to the collision containing the high mass dijet event (note that this is a separate issue to the trigger problem described in Sect. 5.1.1). This issue occurred predominantly for high mass dijet events, and caused their energy to be under-estimated. Since the measurement of the energy deposited in the calorimeter spans several bunch crossings, the correct jet energies could be restored by assigning the event to the correct bunch crossing offline. This procedure was performed manually by experts, and led to an average upward shift of approximately 40% (100%) for the EM (hadronic) jet energy of the affected events. Since the tracking information is recorded within a single bunch crossing, this could not be restored, and hence, for these selected events the recorded tracks are those belonging to the collision in the previous bunch crossing. The overall impact of potential mis-calibration of these events was assessed and was found to be negligible, with respect to the statistical uncertainty on the background estimate. For full details about this issue and its resolution, see [3].

5.1.2.1 Simulated Samples

Descriptions of Monte Carlo simulation techniques were given in Sect. 2.2.3. Unless otherwise stated, the Monte Carlo samples utilised in the high mass dijet analysis were generated at reconstructed level,[1] using Pythia 8 [4] to calculate leading order matrix elements, perform the parton showering, hadronisation and simulation of the underlying event, in conjunction with the A14 set of tuned parameters [5], and the EvtGen program [6] for the decays of beauty and charm hadrons. The NNPDF 2.3 leading order parton distribution functions [7] were used, and the renormalisation and factorisation scales [8] were set to the average transverse momentum of the two leading jets.

In the high mass dijet analysis QCD dijet Monte Carlo is utilised to model the background, with additional jets being provided by the parton shower. An additional truth level QCD dijet Monte Carlo sample was generated, containing more than 2 billion events. Note that the reconstructed level samples were used, unless explicitly stated otherwise. Recall that the background samples are not utilised to derive the background estimate in the analyses described in this thesis. They are utilised in the analysis optimisation, testing of the background estimation procedure, and data-Monte Carlo comparison of variables.

---

[1] In reconstructed level Monte Carlo the simulated particles and additional particles from pile-up are put through a simulation of the ATLAS detector.

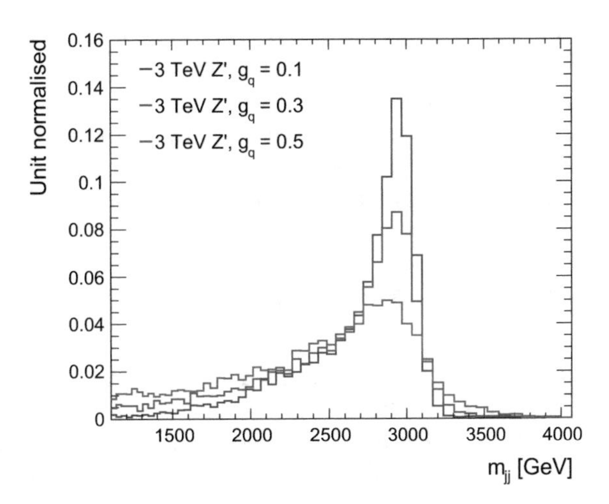

Fig. 5.2 A comparison of the shape of a 3 TeV simulated Z' resonance in the high mass dijet analysis, for a coupling to quarks g_q value of 0.1 (shown in blue), 0.3 (shown in red), and 0.5 (shown in pink). Note that the areas of the signal templates have been normalised to unity, allowing for a comparison of their shapes, but not their normalisation

The 'benchmark' BSM models considered in the analyses in this thesis were introduced in Sect. 2.3. Recall that the Z' dark matter mediator model is utilised in all three analyses, and the other models are only utilised in the high mass dijet analysis. The simulated signals are used in the optimisation of the analysis selection, testing of the background estimation procedure, and in the limit setting.

The leading order matrix elements for the Z' signal samples are calculated using MADGRAPH5 [9]. Only decays to $q\bar{q}$ pairs are simulated for all quark flavours, except for top and bottom quarks. The mass of the dark matter particle m_χ was chosen to be much larger than the mass of the Z' dark matter mediator $m_{Z'}$ such that the Z' does not decay to dark matter. This choice was made to simplify the scenario, and samples were produced for different values of $m_{Z'}$ and g_q only. The width of the Z' increases as the coupling to quarks g_q increases, as illustrated in Fig. 5.2 for a 3 TeV Z' in the high mass dijet analysis. For a quantitative description of the dependence of the width on g_q see [10].

Opposite sign spin 1 W' signals are simulated with coupling values set to the Standard Model W boson couplings, and the coupling of the W' bosons to $W^\pm Z$ is forbidden. Only decays to $q\bar{q}$ are simulated for all six quark flavours. The branching ratio to dijets is 75%, making this the dominant decay mode.

The simulated excited quarks are spin $\frac{1}{2}$ with Standard Model quark couplings, and the compositeness scale[2] is set equal to the mass of the excited quark. Only decays via gluon emission to a qg final state are simulated for up and down quarks. The branching ratio to dijets is 85%, making this the dominant decay mode.

Simulated samples for three models of quantum black holes were produced: two of the models utilise ADD type extra dimensions and are generated by the BLACKMAX [12] and QBH [13] generators, respectively, and the third model utilises RS type extra dimensions (variant RS1) and is generated by the QBH generator.

[2]The compositeness scale is the energy scale above which quark substructure effects could be observed; in the Standard Model this scale is infinite as quarks are assumed to be point-like [11].

For a review about the differences between the BLACKMAX and QBH generators see Appendix A of [14, 15]. The CTEQ6L1 leading order PDF set [16] was utilised and parton showering, hadronisation and underlying event simulation was performed using PYTHIA 8, with the A14 tune. The mass threshold for black hole production M_{Th} was set to the (TeV-scale) fundamental scale of gravity M_D. All decays of the quantum black holes are simulated when generating the MC samples.

5.1.3 Analysis Selection

Now that the data inclusion strategy, datasets and simulated samples have been defined, a description of the analysis selection will be given, including how the data are filtered to ensure that only events gathered when the detector was operating well are included, and how events which are characteristic of a dijet resonance are selected.

Data Quality

The first step in the analysis selection is to choose only data which are good quality and were recorded when the detector was operating well; this is to ensure that flawed data do not bias the analysis results. The data are filtered through the application of a *Good Run List* (GRL), which specifies groups of 'good' luminosity blocks (LBs) to be included in the analysis. Physics Data Quality (DQ) flags indicate the condition of the trigger, reconstruction software, and sub-detectors of the ATLAS detector when each LB was recorded. They are used to determine which LBs are good. The DQ flags, in addition to run requirements (e.g. beam energy) are used to define a GRL [17].

In addition to the requirement that only LBs in the GRL are included, there are also event level quality requirements. Events are removed if they show evidence for noise bursts or data corruption in the calorimeters. In addition, events are vetoed if they were recorded during the recovery procedure for the SCT. Incomplete events which do not have information from the full detector are also vetoed (with the exception of the non-standard data described earlier) [18].

Dijet Topology

The next step in the analysis is to select events which are characteristic of dijet events. The first requirement is that there are at least two tracks associated to the primary vertex. A trigger requirement is then applied. This analysis uses the lowest un-prescaled single jet trigger which was available for the full 3.6 fb^{-1} dataset. The trigger used in this analysis is referred to as the HLT_j360 trigger, which indicates that at least one jet with transverse energy E_T above the 360 GeV threshold was required at the HLT level. The jets reconstructed and calibrated in the HLT are referred to as *trigger jets*. The HLT trigger utilises $R = 0.4$ anti-k_t jets which have been reconstructed and calibrated to the particle level, using a calibration that is similar to the offline calibration which was detailed in Sect. 4.2.

This analysis uses $R = 0.4$ anti-k_t jets which are reconstructed offline then calibrated to the particle level in the analysis. Only events with at least two jets which pass jet cleaning criteria and have $p_T > 50$ GeV are included. The p_T threshold of 50 GeV was chosen in order to reduce the effects of pile-up on the analysis. Since the trigger jets which were used to select which events passed the trigger do not use the final jet calibration that is applied to offline jets, their p_T can differ from the offline jet values. In order to ensure that the analysis is not affected by any trigger efficiency bias, the following cuts are applied: leading jet $p_T > 440$ GeV and dijet invariant mass $m_{jj} \geq 1.1$ TeV GeV. Details about the selection of these requirements and additional cross-checks performed will be given in the following sections.

Jet Cleaning

The cleaning criteria used in this analysis are recommended by ATLAS, and full details are provided in Ref. [19]. The goal of the jet cleaning criteria are to remove 'fake' jets, while retaining a high selection efficiency for the 'real' jets originating from the hard scattering process. The sources of fake jets targeted by the jet cleaning are summarised below, together with their characteristic features which are used to identify them for removal.

The data quality checks previously mentioned remove most of the events which display a large amount of noise in the calorimeters; however, some noise still remains and is removed through the application of jet cleaning. One source of calorimeter noise is from noise bursts in the Hadronic End Cap (HEC) calorimeter, which leads to individual noisy calorimeter cells. Jets reconstructed from these noisy cells usually possess the following features:

- The fraction of their total energy from deposits in the HEC calorimeter f_{HEC} is large.
- The average signal shape quality $\langle Q \rangle$ is large for these jets, meaning there is a poor match between the observed calorimeter signal shape and one expected for a real jet.
- The fraction of their energy in HEC calorimeter cells with a poor signal shape quality f_Q^{HEC} is high.
- The total energy of all cells with an apparent negative energy E_{neg} is large for these jets.

In addition to sporadic noise bursts in the HEC, coherent calorimeter noise across many cells in the EM calorimeter can occur. Jets reconstructed from these cells usually possess the following features:

- The fraction of the total energy of the jet from deposits in the EM calorimeter f_{EM} is high.
- The average signal shape quality $\langle Q \rangle$ is large for these jets.
- The fraction of their energy in the LAr calorimeter cells with a poor signal shape quality f_Q^{LAr} is high.

In addition to jets arising from calorimeter noise, jet cleaning also targets jets produced by Non-Collision Backgrounds (NCB). This covers two sources of fake jets.

One of these sources is beam induced background, where protons which are lost from the beam upstream can interact with the collimators or other shielding, resulting in cascades of particles. Additionally, protons can interact with any residual gas in the beam pipe, leading to particle cascades. High-energy muons produced in these cascades are capable of passing through the shielding and leave energy deposits in the detector which can be reconstructed as jets [20].

The other source of NCB is from cosmic-rays, where showers of particles produced in the atmosphere can lead to additional particles passing through the ATLAS detector. The particles which reach ATLAS are mainly muons, as it is deep underground.

Jets arising from NCB sources are distinguished from real jets in the following ways:

- Real jets usually contain charged hadrons and originate from the interaction point, depositing some energy in the EM calorimeter, so a minimum requirement on f_{EM} is made. Additionally, since charged hadrons produce tracks, a minimum requirement on the jet charged fraction f_{ch} is made, where f_{ch} is the ratio of the scalar p_T sum of tracks from the primary vertex associated to the jet, divided by the jet p_T.
- Real jets tend to extend from the interaction point, depositing energy in many calorimeter layers rather than being localised, which is more typical of jets from NCB. The maximum energy fraction in a single calorimeter layer f_{max} being high is an indicator of the jet arising from NCB, or from calorimeter noise.

A summary of the variables used to discriminate between fake jets and real jets has been given, and the exact requirements on the variables mentioned is provided in Appendix A. The recommended jet cleaning criteria were largely based on the previous studies described in [20, 21], with additional optimisations performed using 6.4 pb^{-1} $\sqrt{s} = 13$ TeV data. Two recommendations were provided, with the loose working point having efficiencies of 99.5% (99.9%) for $p_T > 20\,(100)$ GeV, and the tight working point having efficiencies of 95% (99.5%) for $p_T > 20\,(100)$ GeV. The loose working point was chosen for this analysis.

The jet cleaning criteria were derived using distributions which were inclusive in jet p_T, and only a small $\sqrt{s} = 13$ TeV dataset was used in the optimisation, so the high p_T region was not investigated by the cleaning studies. Since the high mass analysis is particularly interested in jets with p_T up to several TeV, it was necessary to check the performance of the cleaning criteria at high p_T. Many cross-checks were performed, one of which was investigating jet timing as a function of η, where jet timing is given by the energy-squared weighted average of the calorimeter cell time, for all cells belonging to the jet. If the jets are coming from the hard scatter process, the jet timing should be approximately zero for all η values; from Fig. 5.3 it is seen that this is the case. Figure 5.3 uses the full 3.6 fb^{-1} of data, after application of the full analysis selection given in Sect. 5.1.3, this is also the case for all the other figures shown in this chapter, unless otherwise stated.

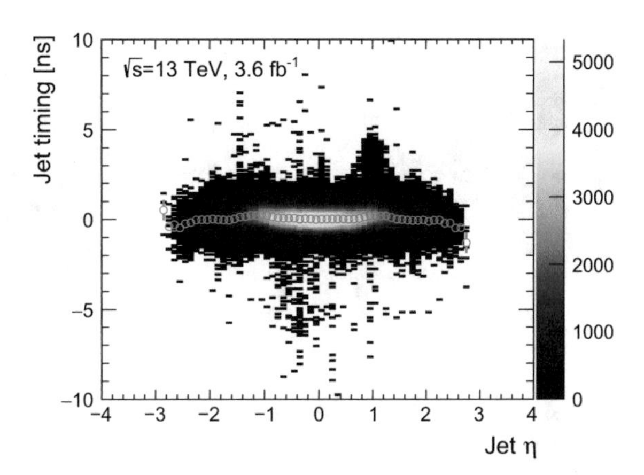

Fig. 5.3 The jet timing versus jet η distribution is shown for the leading two jets in data. Significant deviations from a jet timing of zero is indicative of NCB, noise, or poorly measured jets. The average value of jet timing, indicated by the blue circles, is approximately flat and centered around zero as a function of jet η, indicating that the cleaning criteria are performing well

Tile Calorimeter Module Masking

Two tile calorimeter modules were not operational during Run II data taking and were masked in the reconstruction. One of these modules failed early on and could be simulated in the MC. Since one module was not simulated in the MC, this introduced a small miscalibration for jets falling in this region. This miscalibration is partially corrected by the η-intercalibration.

Pile-up Reduction

In order to reduce the effects of pile-up on the analysis, a minimum p_T requirement of 50 GeV was made on both the leading and sub-leading jets. In Fig. 5.4a, the distribution of jet multiplicity as a function of N_{PV} is shown to be approximately flat for jets with $p_T > 50$ GeV, indicating that additional pile-up jets are being removed effectively.

In addition to introducing extra jets to the event, pile-up can also increase the energy of jets in the event. This effect is mainly dealt with by the jet area-based pile-up correction described in Sect. 4.2.2. As a cross-check, Fig. 5.4b was produced. It shows that m_{jj} as a function of N_{PV} is approximately flat, indicating that additional jet energy due to pile-up is being removed successfully.

Trigger Efficiency

The thresholds for the leading jet p_T cut and the cut on m_{jj} are determined by producing trigger efficiency curves in the analysis phase space, and determining the point at which the efficiency reaches 99.5%. The first step is to apply the analysis selections mentioned so far, in addition to a signal optimisation selection $|y^*_{12}| < 0.6$, detailed in Sect. 5.1.3, such that we are examining data in the phase space of the analysis. The trigger efficiency is calculated bin-by-bin as a function of leading jet p_T and m_{jj} by taking the ratio between an un-biased sample of events which pass a single jet trigger with a lower E_T threshold, the *reference trigger*, and events passing both the reference trigger and the HLT_j360 trigger. The resulting efficiency curves

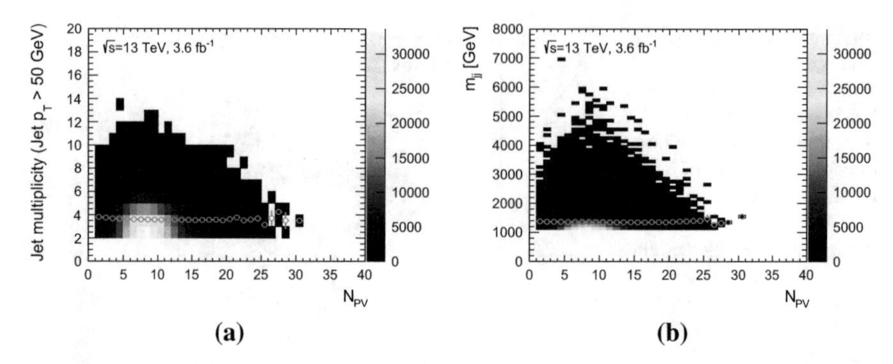

Fig. 5.4 Figure **a** shows jet multiplicity (for jets with $p_T > 50$ GeV) as a function of N_{PV} in data. The mean value of jet multiplicity, indicated by the blue circles, is seen to be flat as a function of N_{PV}, indicating that additional jets from pile-up are being removed effectively. Figure **b** shows m_{jj} as a function of N_{PV} in data. The mean value of m_{jj}, indicated by the blue circles, is seen to be flat as a function of N_{PV}, indicating that that additional energy from pile-up is being removed from the jets which form the dijet

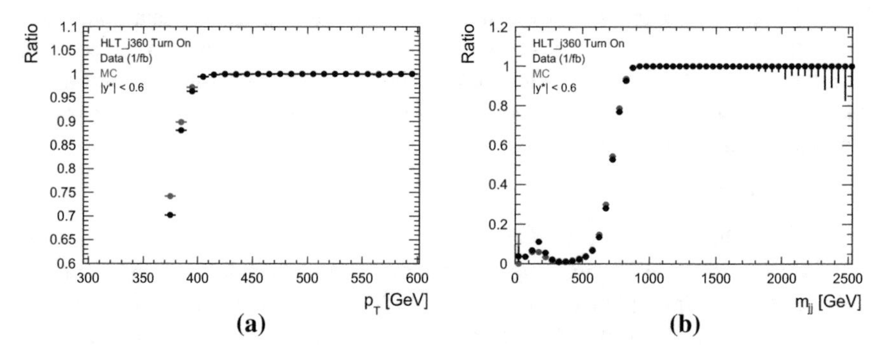

Fig. 5.5 Trigger efficiency curves as a function of **a** leading jet p_T and **b** m_{jj}, are shown for MC in red, and for 1 fb^{-1} of $\sqrt{s} = 13$ TeV data in black. Figure taken from [22]

Table 5.1 This table shows the values where the trigger efficiency plateau (defined as 99.5% efficiency) is reached for leading jet p_T and m_{jj}, and the offline cut values used in the analysis

Variable	Plateau [GeV]	Offline cut [GeV]
Lead jet p_T	409	440
m_{jj}	900	1100

are shown in Fig. 5.5a, b as a function of leading jet p_T and m_{jj}, respectively. A fit to the curves was used to determine the position of the trigger plateau (defined by 99.5% efficiency).

Table 5.1 summarises the points where the plateau is reached, together with the values chosen for the cuts in the analysis. It should be noted that the cut values chosen by the analysis are conservative. One reason for this is that the analysis cuts were set

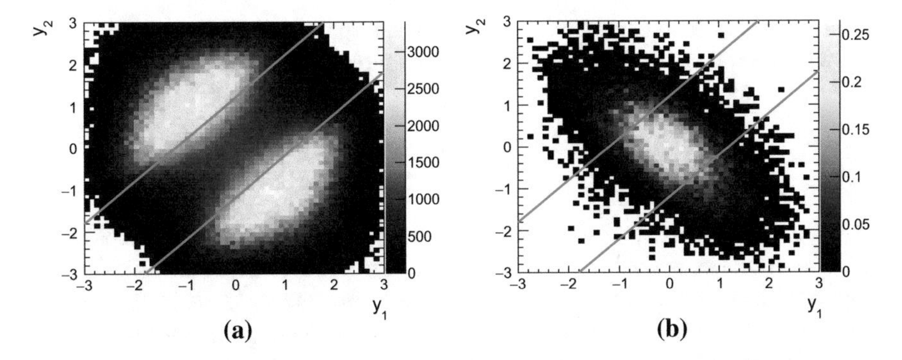

Fig. 5.6 The rapidity for the sub-leading jet versus the rapidity for the leading jet is shown for **a** SM QCD processes, and **b** a 5 TeV q*, produced using Monte Carlo. The full analysis selection has been applied (with the exception of the y^*_{12} cut), hence, these distributions are for $m_{jj} \geq 1.1$ TeV. The blue lines indicate the $|y^*_{12}| = 0.6$ position

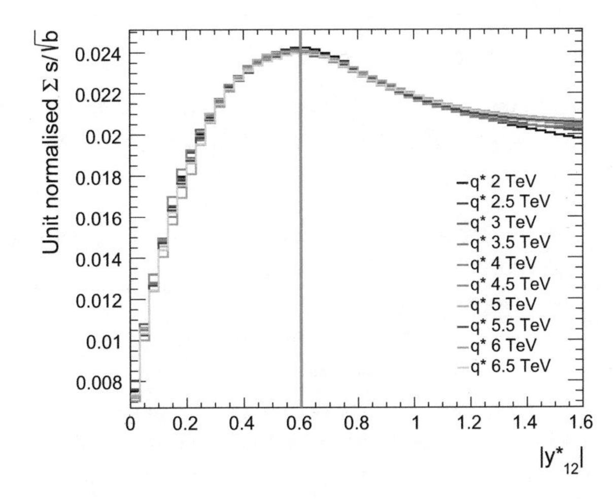

Fig. 5.7 The signal significance $\frac{s}{\sqrt{b}}$ integrated to the $|y^*_{12}|$ cut position is shown for the excited quark model, for a range of signal masses. Note that each of the significance histograms has been normalised to unit area over the displayed region. The position that maximises the integrated $\frac{s}{\sqrt{b}}$ is $|y^*_{12}| = 0.6$, and is indicated by the blue line

before data-taking, so by giving a larger margin between the plateau and the offline cut, the offline cut values would still be suitable if the lowest un-prescaled jet trigger threshold was increased during data-taking.

Signal Optimisation

In order to enhance our sensitivity to BSM models and reject QCD background, a requirement is made on the rapidity of the jets which form the dijet. As mentioned in Sect. 2.2.3, QCD background is mainly the result of t-channel processes resulting in the production of jets at small scattering angles. BSM physics is produced via s-channel processes resulting in a more isotropic or central distribution of jets. These features are illustrated in Fig. 5.6, which shows the rapidity of the sub-leading jet versus the rapidity of the leading jet for SM QCD processes, and for a BSM model (5 TeV excited quark).

In order to reject the forward peaking QCD background, the absolute value of the y^*_{12} variable was used, which is related to the rapidity separation between the two leading jets by

$$|y^*_{12}| = \frac{|y_1 - y_2|}{2}, \tag{5.3}$$

where y_1 is the rapidity of the leading jet and y_2 is the rapidity of the sub-leading jet. The optimum cut value for this variable was determined by investigating the impact of the cut value on the signal significance $\frac{s}{\sqrt{b}}$, where s is the total number of signal events kept by the selection and b is the corresponding value for background events. Figure 5.7 shows the signal significance integrated to the cut position for a range of cut values, showing that the optimum value (position that maximises the integrated $\frac{s}{\sqrt{b}}$) for the excited quark model is $|y^*_{12}| = 0.6$. The analysis selection therefore requires that events have $|y^*_{12}| < 0.6$, the lines indicating $|y^*_{12}| = 0.6$ are shown on Fig. 5.6.

Summary of Analysis Selection

A summary of the final analysis selection is given in Table 5.2. The event yields at each stage of this selection are given in Appendix C.

Table 5.2 The analysis selection used in the high mass dijet analysis

High mass dijet analysis selection
Remove events which:
– Do not belong to good LBs, defined in the Good Run List;
– Show evidence for noise bursts;
– Show evidence for data corruption in the calorimeters;
– Were recorded during the recovery procedure for the SCT;
– Are incomplete, i.e. they do not have information from the full detector.
Events must:
– Have a primary vertex with at least two associated tracks;
– Contain at least two clean jets with $p_T > 50$ GeV;
– Pass HLT_j360 trigger;
– Contain a jet with $p_T > 440$ GeV.
Two leading jets must satisfy:
– $|y^*_{12}| < 0.6$;
– $m_{jj} \geq 1.1$ TeV.

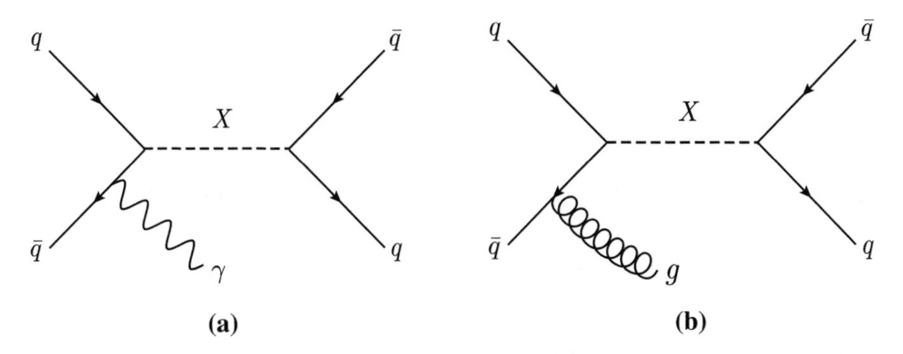

Fig. 5.8 Example Feynman diagrams showing the production of a new resonance X decaying to a dijet, and produced in association with **a** an ISR photon and **b** an ISR jet

5.2 The Dijet + Initial State Radiation Analyses

Recall that the dijet + ISR analyses target the low dijet mass region. This region is particularly challenging due to the high production rate of jets coupled with limitations in data recording rates, forcing us to introduce high pre-scales on jet triggers with low E_T requirements. As this region is difficult to access, some of the most stringent limits are from older experiments with lower centre-of-mass energies, using small datasets. Therefore, this region is particularly interesting to explore with larger datasets if there is a way to efficiently trigger on low mass dijet events.

The dijet + ISR analyses presented in this thesis use an innovative approach [23, 24] to help overcome the trigger limitations. A high momentum object (a photon or a jet) radiated from the initial state could recoil against a light resonance, boosting it in the transverse direction. The ISR object is used to trigger the event, allowing us to gather low mass dijet events more efficiently than if we were to trigger on the decay products of the resonance directly. See Fig. 5.8 for an example Feynman diagram for each process.

These analyses were performed for the first time ever in ATLAS in 2016. A conference note for the dijet + γ analysis was produced for the LHCP conference in June 2016 [25], using 3.2 fb^{-1} of pp collision data. These results were superseded and a conference note was produced for the ICHEP conference in August 2016 [26], using 15.5 fb^{-1} of pp collision data. Only the ICHEP analysis will be described in this thesis. Results from both the dijet + γ analysis and the dijet + jet analysis were included in this result, and the same un-blinded approach utilised in the high mass dijet analysis was followed.

5.2.1 Datasets and Simulated Samples

The dijet + ISR analyses utilise 15.5 fb^{-1} of $\sqrt{s} = 13$ TeV data collected in 2015 and 2016. The number of events in the debug stream was found to be negligible in the phase space of the analysis, hence, these events were not included. All of the data included in the analysis was collected when the full detector was operating well. The data recorded with 50 ns bunch spacing, and when the IBL was off or in standby mode was not included. Additionally, no events assigned to the wrong bunch crossing were found after the application of the analysis selections. The simulated background samples for the dijet + jet analysis are the same as those utilised in the high mass dijet analysis.

For the dijet + γ analysis, background samples were simulated using SHERPA 2 [27] to calculate leading order matrix elements in α and α_s for up to three or four partons (three for events with a photon p_T of less than 70 GeV, and four for events with higher photon p_T). SHERPA 2 was also utilised to perform the parton showering, hadronisation and simulation of the underlying event. To avoid overlap between the emissions from the matrix element and the emissions from the parton shower, the two are merged using the leading order prescription described in [28]. The CT10 next-to-leading order parton distribution functions [29] were utilised.

The signal samples utilised in the dijet + ISR analyses are the same as the Z' dark matter mediator samples described for the high mass dijet analysis; however, decays to beauty quarks are also included, and different Z' masses are considered.

An additional difference between the MC samples utilised in the dijet + ISR analyses and those utilised in the high mass dijet analysis is that pile-up re-weighting has been applied to the dijet + ISR reconstructed level MC samples. Recall that μ is the mean number of simultaneous inelastic proton-proton interactions being recorded in a bunch crossing; pile-up re-weighting means that the MC events are weighted such that the μ distribution in MC matches the μ distribution in data [30], shown in Fig. 3.2a. In addition, for MC samples utilised in the dijet + γ analysis, correction factors were applied to photons in the MC to obtain better agreement between shower shape variables in data and MC, as described in [31].

5.2.2 *Mass Fractions*

As previously mentioned, each analysis needs to determine which pair of jets is most likely to have been produced by the decay of the resonance in order to calculate the dijet invariant mass. For the dijet + γ analysis, the γ is assumed to be the ISR object and the invariant mass m_{jj} is calculated using the leading and sub-leading jets. For the dijet + jet analysis, the situation is more complex since all three final state objects are of the same type. Therefore, in principle one could use m_{12}, m_{13} or m_{23} as the observable, where the numbers denote which p_T ordered jets are used to calculate the mass, with 1 being the leading jet.

In order to determine which mass combination is most suitable for reconstructing the resonance, plots showing the fraction of 'correct pairing' versus nominal signal mass were made for each mass combination. In order to produce these plots, the number of events in a $\pm 10\%$ mass window around the nominal signal mass is calculated for each mass combination, and then converted into a fraction by dividing by the

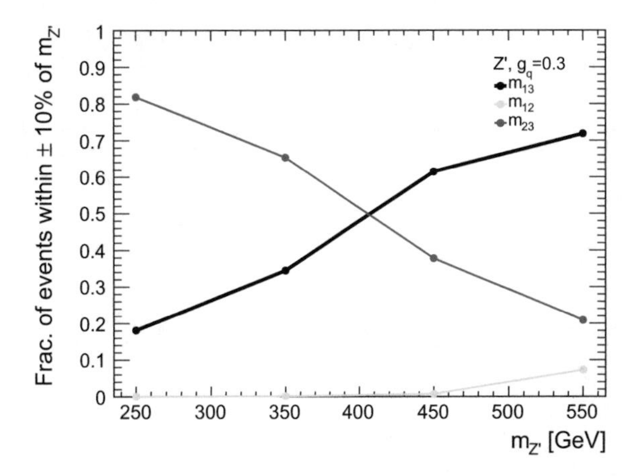

Fig. 5.9 The fraction of events which are reconstructed in a $\pm 10\%$ mass window around the nominal signal mass, for a Z' signal with $g_q = 0.3$, as a function of nominal signal mass. The fractions are shown for each mass combination, m_{12}, m_{13} and m_{23}. The fractions are computed relative to the total number of events within the $\pm 10\%$ mass window around the nominal signal mass for all three mass combinations

total number of events within the window for all three mass combinations. Figure 5.9 shows an example of one of the plots that are produced. This figure uses the Z' signal with $g_q = 0.3$, but similar features are seen for the other g_q couplings. Note that the final analysis selection, outlined in Sect. 5.2.3, was applied to the signals prior to calculating the fractions and the y^* cut applied is between the signal jets (e.g $|y^*_{13}|$ is used in the selection for m_{13}).

The figure shows that m_{23} is the optimum choice in the low mass region (up to $\sim$400 GeV), above this threshold m_{13} becomes optimum. As this analysis is interested in accessing the low mass region, m_{23} was chosen as our observable, referred to as m_{jj} in the following. In future analyses, one could consider additionally performing a search using m_{13}, but for simplicity, and since this was a new analysis in ATLAS, it was decided not to pursue this.

5.2.3 Summary of Analysis Selections

The dijet + ISR analyses share many similar analysis selections to the high mass dijet analysis described in Sect. 5.1.3. A summary of the jet and photon requirements for the dijet + ISR analyses is given in Table 5.3, before giving the final analysis selections in Table 5.4. The event yields at each stage of the analysis selections are provided in Appendix C.

Descriptions of requirements that differ to those used in the high mass analysis will now be given. Note that the event level data quality requirements are the same

Table 5.3 The jet and photon object selection used in the dijet + ISR analyses

Dijet + ISR object selection	
Jet requirements	Photon requirements
Use clean $R = 0.4$ anti-k_t jets; within $\|\eta\| < 2.8$; with $p_T > 25\,\text{GeV}$; and JVT > 0.59 for $p_T < 60\,\text{GeV}$ & $\|\eta\| < 2.4$.	Use converted and unconverted, isolated γ; Satisfying tight identification; with $\|\eta\| < 2.37$; excluding $1.37 < \|\eta\| < 1.52$.

Table 5.4 The analysis selections used in the dijet + ISR analyses

Dijet + ISR analysis selections	
Remove events which:	
– Do not belong to good LBs, defined in the Good Run List;	
– Show evidence for noise bursts;	
– Show evidence for data corruption in the calorimeters;	
– Were recorded during the recovery procedure for the SCT;	
– Are incomplete, i.e. they do not have information from the full detector.	
The primary vertex must have at least two tracks associated with it.	
Dijet + γ	*Dijet* + jet
Events must:	Events must:
– Contain at least two clean jets;	– Contain at least three clean jets;
– Pass HLT_g140_loose trigger;	– Pass HLT_j380 trigger;
– Contain a γ with $p_T > 150\,\text{GeV}$.	– Contain a jet with $p_T > 430\,\text{GeV}$.
Two leading jets must satisfy:	Second and third jets must satisfy:
– $\|y^*_{12}\| < 0.8$;	– $\|y^*_{23}\| < 0.6$;
– $169 \leq m_{jj} \leq 1493\,\text{GeV}$;	– $303 \leq m_{jj} \leq 611\,\text{GeV}$.
– $\Delta R_{\text{ISR,close-jet}} > 0.85$.	
Overlap removal is applied.	

as those used in the high mass dijet analysis, except for the application of a different GRL for this higher luminosity dataset. Hence, these requirements will not be mentioned further.

Jet Requirements

The reduction of the jet p_T requirement to 25 GeV allows us to explore lower dijet masses, but introduces the need to employ additional pile-up mitigation techniques. In the jet p_T region below 60 GeV, the Jet Vertex Tagger (JVT) [32] is used in order to distinguish jets originating from the primary vertex from pile-up jets. This is a multivariate discriminant which uses tracking information, hence, its range is limited to jets within $|\eta| < 2.4$. The default cut value of 0.59 is used [32]. After the application of the JVT requirement and other pile-up mitigation techniques described

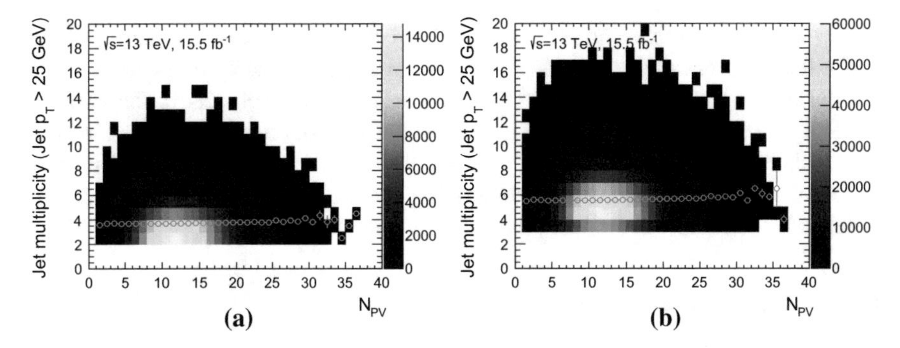

Fig. 5.10 The jet multiplicity (for jets with $p_T > 25\,\text{GeV}$) as a function of N_{PV} in data is shown for **a** the dijet + γ analysis, and **b** the dijet + jet analysis. In both cases the mean value of jet multiplicity, indicated by the blue circles, is seen to be flat as a function of N_{PV}, indicating that additional jets from pile-up are being removed effectively

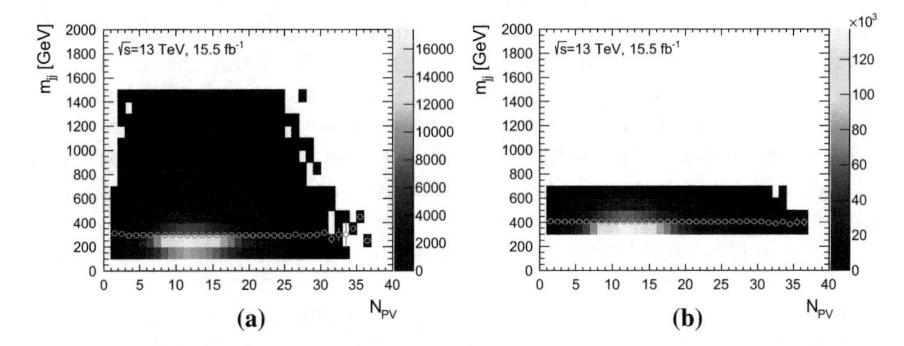

Fig. 5.11 Dijet invariant mass m_{jj} as a function of N_{PV} in data is shown for **a** the dijet + γ analysis, and **b** the dijet + jet analysis. In both cases, the mean value of m_{jj}, indicated by the blue circles, is seen to be flat as a function of N_{PV}, indicating that the additional energy from pile-up is being removed from the jets which form the dijet

earlier, such as the jet area-based pile-up correction, pile-up has a negligible effect, as demonstrated in Figs. 5.10 and 5.11.

Photon Requirements

Both converted and unconverted photons are used in this analysis, and they are reconstructed as described in Sect. 4.4 and calibrated as described in [33, 34]. In order to select events with a prompt photon, rather than those arising from hadronic decays, or the false identification of an electron or a jet as a photon, the photon is required to satisfy *tight identification*. This identification procedure uses requirements on shower shape information from the EM calorimeter, the energy fraction deposited in the hadronic calorimeter and information from the finely segmented first layer of the EM calorimeter [35]. The photon is required to be within $|\eta| < 2.37$, as this is the region spanned by the finely segmented first layer of the EM calorimeter. Within this region, photons within $1.37 < |\eta| < 1.52$ are excluded, as photons are poorly measured in this transition region between the barrel and end-cap. In order to further suppress the selection of photons arising from background sources, the photon is required to be isolated, i.e. separated from other objects in the event. A cone of size $\Delta R = \sqrt{(\eta - \eta^{\gamma})^2 + (\phi - \phi^{\gamma})^2} = 0.4$ is defined around the direction of the photon, and the sum of E_T of the topoclusters in this cone (after subtraction of the photon energy, pile-up and underlying event contributions) is calculated. In order to be classified as isolated, the sum E_T must be less than 2.88 GeV + 0.024 p_T^{γ} [36].

Dijet + ISR Analysis Selections

In the dijet + γ analysis, the lowest un-prescaled single photon trigger available was used. The trigger is referred to as the HLT_g140_loose trigger, which indicates that at least one photon with E_T above the 140 GeV threshold was required at the HLT level. The photon was required to satisfy loose identification at the trigger level. The HLT trigger uses photons which have been reconstructed and calibrated in the HLT, using a calibration that is similar to the offline calibration. The leading photon in this analysis is required to have a p_T above 150 GeV, in order to be above the trigger plateau. Mass cuts are applied for these analyses, but unlike the high mass dijet analysis, the range is not set directly by deriving the trigger turn on in m_{jj}. The reason for this is that m_{jj} is calculated independently of the leading photon on which we trigger; it is calculated using the two highest p_T jets. The values of the mass cuts that we apply are based on the region over which we fit the spectrum in order to derive the background estimate; more details about the choice of this range will be given in Sect. 6.2.3. Overlap removal is applied as a further isolation requirement within the isolation cone used in the analysis, and to address the problem of duplication, i.e. the reconstruction of a single object as two separate objects [37]. In this analysis overlap removal is applied between the photon and the jet, discarding jets which are within $\Delta R = \sqrt{(y - y^{\gamma})^2 + (\phi - \phi^{\gamma})^2} = 0.4$ around the photon, as recommended in [37]. The combined application of the previous requirements selects a photon sample with greater than 90% (80%) purity for unconverted (converted) photon candidates [38].

In the dijet + jet analysis the HLT_j380 trigger is used, as the E_T threshold for the lowest un-prescaled single jet trigger was raised from 360 to 380 GeV during 2016. A leading jet p_T cut of 430 GeV was utilised. This is lower than the leading jet p_T cut applied in the high mass dijet analysis, even though the trigger threshold has been raised. The value used in the high mass analysis was very conservative to ensure that if the trigger threshold was raised during data taking then the leading jet p_T cut could remain unchanged. Since we are interested in the low mass region now, it is beneficial to lower this requirement as much as possible. Mass cuts are applied in this analysis. As for the dijet + γ case, the values for these cuts are based on the range over which we fit the spectrum; more details about the choice of this range will be given in Sect. 6.2.3.

Signal Optimisation

In order to optimise our sensitivity to BSM models, additional analysis cuts were applied to preferentially select signal events while rejecting background events. The

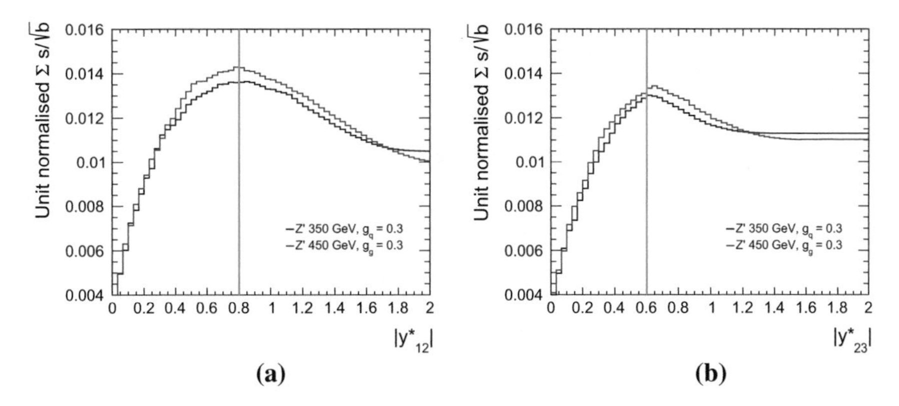

Fig. 5.12 The signal significance $\frac{s}{\sqrt{b}}$ integrated to the cut position is shown for the Z' model with $g_q = 0.3$, for two signal masses. Note that each of the significance histograms has been normalised to unit area over the displayed region. The position that maximises the integrated $\frac{s}{\sqrt{b}}$ is **a** $|y^*_{12}| = 0.8$ for the dijet + γ analysis and **b** $|y^*_{23}| = 0.6$ for the dijet + jet analysis, and is indicated in the figures by the blue line

same technique as outlined in Sect. 5.1.3 was utilised in order to decide the optimal cut positions. Two signal points for the Z' model were considered in the optimisation, with masses of 350 and 450 GeV and coupling $g_q = 0.3$. For the optimisation, a mass window cut of ± 50 GeV around the nominal signal mass was applied to m_{12} for the dijet + γ analysis, and to m_{23} for the dijet + jet analysis. Utilising a mass window cut ensures that we are performing the optimisation using events for which we selected the correct mass combination to reconstruct the resonance peak, rather than performing the optimisation using 'wrong combination' events. The cut is applied to both signal and background samples to ensure we are comparing events in the same mass region. The window size utilised roughly corresponds to 3σ of the width of the Z' mass points with $g_q = 0.3$.

In order to reject the forward peaking QCD background, the absolute value of the y^*_{12} variable was used in the dijet + γ analysis, as defined previously in Eq. (5.3). In the dijet + jet analysis, the absolute value of the y^*_{23} variable was used. This variable is defined by

$$|y^*_{23}| = \frac{|y_2 - y_3|}{2}, \tag{5.4}$$

where y_2 is the rapidity of the sub-leading jet and y_3 is the rapidity of the third highest p_T jet in the event. In each of the two analyses, the variables relate to the rapidity separation between the 'signal jets', i.e. the jets used to calculate the dijet invaraint mass. Figure 5.12 indicates that $|y^*_{12}| < 0.8$ is the optimum selection for the dijet + γ analysis, and $|y^*_{23}| < 0.6$ is the optimum selection for the dijet + jet analysis.

For the dijet + γ analysis, a slight gain is achieved by applying an additional requirement that the leading photon in the event is separated from the closest jet by

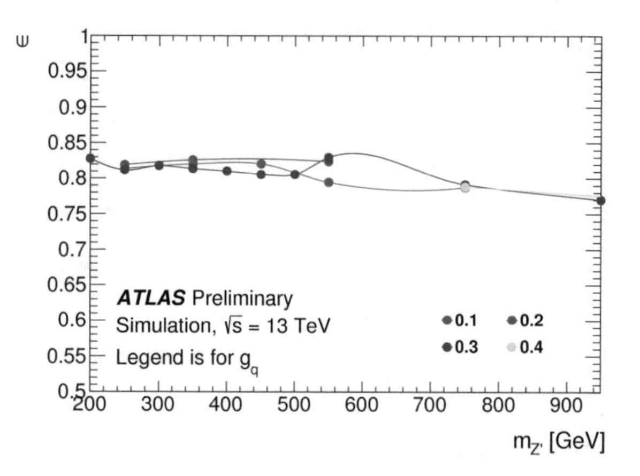

Fig. 5.13 A summary of the efficiencies for all the mass and coupling points utilised in the dijet + γ analysis. Figure taken from [26]

$\Delta R_{\text{ISR,close-jet}} > 0.85$. This requirement rejects events in which the photon is close to or inside a jet.

The efficiency ϵ of the dijet + ISR selections were evaluated for each signal mass and coupling point considered in the analyses. Efficiency is defined as the fraction of events passing the analysis selection in reconstructed level Monte Carlo with respect to the number of events passing the analysis selection in truth level Monte Carlo. This takes into account the photon and jet reconstruction and identification efficiencies. The efficiencies for the dijet + γ analysis are summarised in Fig. 5.13, with an average efficiency of 81%. For the dijet + jet analysis (and the high mass dijet analysis) the efficiency is 100%.

5.3 Data-Monte Carlo Comparisons

Prior to producing the dijet invariant mass spectra, an important cross-check is to compare the data to MC predictions in the phase space of the analysis and look for any discrepancies. A small sample of data-MC comparisons for some of the key analysis variables are shown in Fig. 5.14 for the high mass dijet analysis, Fig. 5.15 for the dijet + γ analysis, and Fig. 5.16 for the dijet + jet analysis.

For the high mass dijet analysis comparisons a normalisation scale factor of 0.87 was applied to the MC, as the MC cross-section was found to be ~13% too high with respect to data. For the dijet + γ case a normalisation scale factor of 1.52 was applied to the MC, and for the dijet + jet case a normalised scale factor of 0.83 was applied to the MC. These figures show that in general the data and MC are consistent within the jet energy scale uncertainty bands (jet energy scale and photon energy scale uncertainties for the dijet + γ analysis).

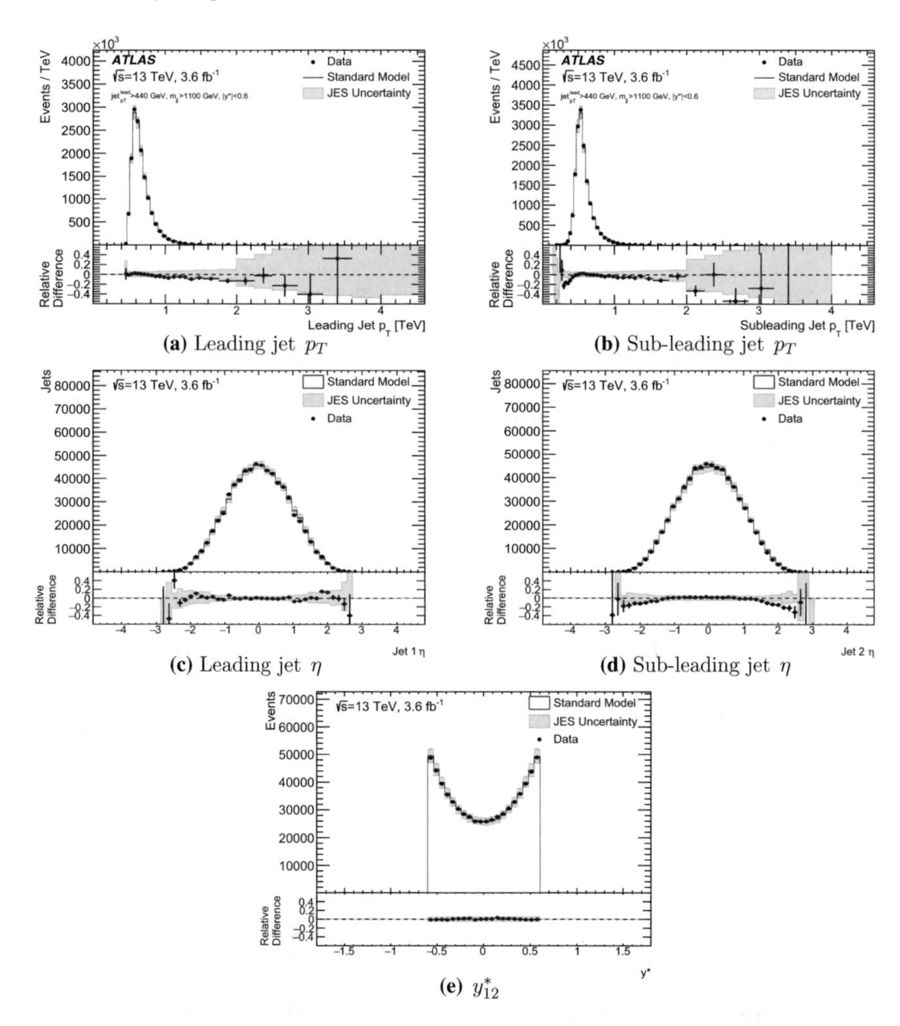

(a) Leading jet p_T **(b)** Sub-leading jet p_T

(c) Leading jet η **(d)** Sub-leading jet η

(e) y^*_{12}

Fig. 5.14 A data-MC comparison of some of the key variables used in the high mass dijet analysis. The full 3.6 fb^{-1} dataset is compared to MC generated using PYTHIA and scaled by a factor of 0.87. The blue shaded bands show the JES uncertainties. Figures **a** and **b** are taken from [39]. Figures **c**, **d** and **e** are taken from [22]

5.4 Producing the Spectra

Once appropriate cross-checks have been made, the next step in each analysis is to produce a dijet invariant mass spectrum. In order to do this, we must first select an appropriate binning for the spectrum. The binning utilised in the high mass dijet analysis will first be described, before describing the binning utilised in the dijet + ISR analyses.

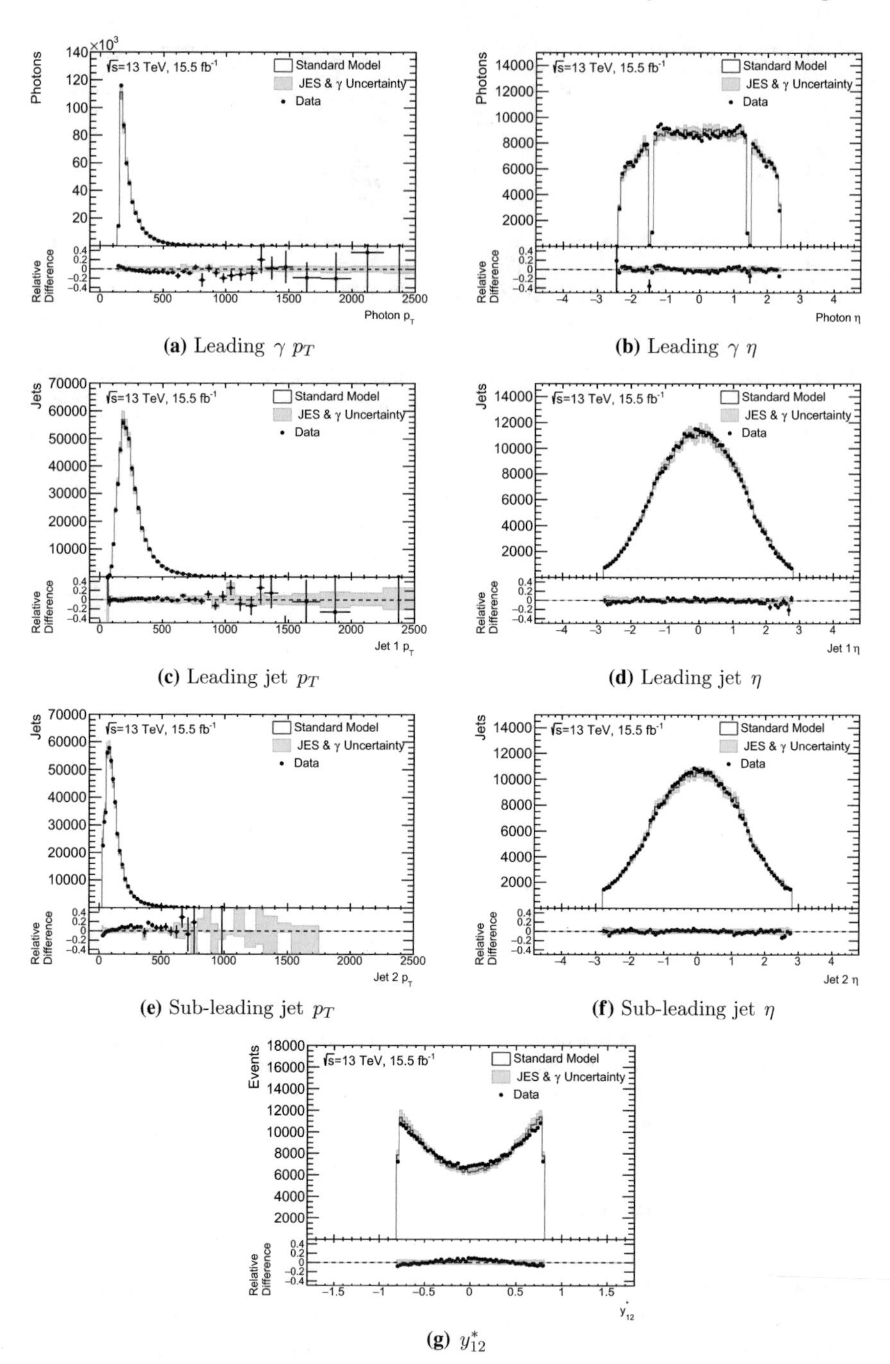

Fig. 5.15 A data-MC comparison of some of the key variables used in the dijet + γ analysis. The full 15.5 fb^{-1} dataset is compared to MC generated using SHERPA and scaled by a factor of 1.52. The blue shaded bands show the combination of the jet energy scale and photon energy scale uncertainties

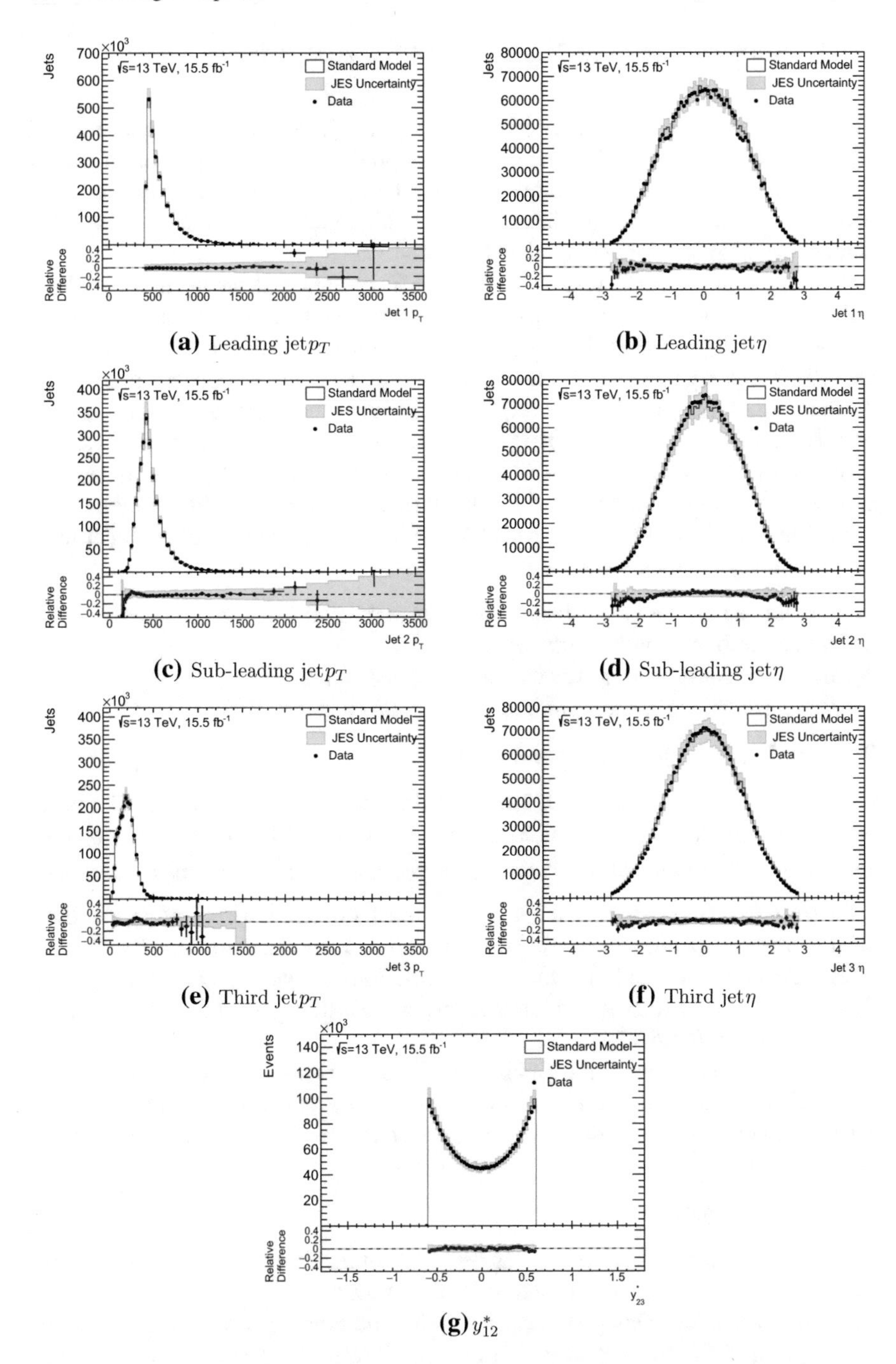

(a) Leading jetp_T **(b)** Leading jetη

(c) Sub-leading jetp_T **(d)** Sub-leading jetη

(e) Third jetp_T **(f)** Third jetη

(g) y^*_{12}

Fig. 5.16 A data-MC comparison of some of the key variables used in the dijet + jet analysis. The full 15.5 fb^{-1} dataset is compared to MC generated using PYTHIA and scaled by a factor of 0.83. The blue shaded bands show the jet energy scale uncertainties

5.4.1 Spectrum for the High Mass Dijet Analysis

There are several criteria that the derived binning should satisfy. The selected binning should be wider than the dijet mass resolution to decrease the migration of events from one invariant mass bin to another, but narrower than our expected signal width, such that signals span several bins, rather than falling into one single bin. Narrower binning can help to improve our signal sensitivity.

In order to search for excesses above the QCD background, we need to be able to estimate this background. The approach used in the high mass dijet analysis is to fit the spectrum using a smoothly falling function in order to obtain the background estimate. This strategy impacts our choice of binning as our spectrum needs to preserve the smoothly falling nature of the QCD background, not introducing false bumps or kinks to the spectrum through a poor choice of binning, as sharp features in the spectrum can not be described by our function. Additionally, it should be noted that narrower bins provide more inputs for the fit function, improving fit stability.

Taking all of the factors mentioned above into account, the following set of criteria for the binning selection were devised:

1. Bin widths must be larger than the dijet mass resolution.
2. Number of bins should be maximised.
3. Bin widths must be narrower than the expected signal width.
4. Binning should produce a smooth spectrum when using QCD dijet MC.

Determining the Dijet Mass Resolution

In order to set a lower bound on the bin widths, we need to derive the dijet mass resolution. The dijet mass resolution is derived using QCD dijet MC and calculating the ratio of $\frac{m_{jj}^{\mathrm{reco}}}{m_{jj}^{\mathrm{truth}}}$ in bins of m_{jj}^{truth}, where the reconstructed jets are ghost associated to the truth jets. A Gaussian shape is fitted to the distribution in each m_{jj}^{truth} bin, then the dijet mass resolution is calculated as the width of the Gaussian fit, $\sigma(m_{jj})$. This is translated into the fractional dijet mass resolution by dividing by the mean of the Gaussian $\langle m_{jj} \rangle$. The fractional dijet mass resolution is then plotted against truth level m_{jj}, as shown in Fig. 5.17.

Note that when deriving the binning, the final high mass dijet analysis selection was applied, with the exception of the leading jet p_T cut, which was loosened to 350 GeV, in order to avoid any kinematic bias on the resolution at the low mass end of the spectrum.

Deriving the Binning

Using the criteria stated earlier, it was decided that maximising the number of bins was the optimum choice for this analysis. This translates to picking bin widths that match the resolution. Our first step in deriving the binning is to fit the histogram shown in Fig. 5.17 with a polynomial function in order to derive a smooth curve. Above 8055 GeV, there were insufficient MC events to derive the resolution, so a flat extrapolation was used to extend the curve. This is a reasonable approximation as in

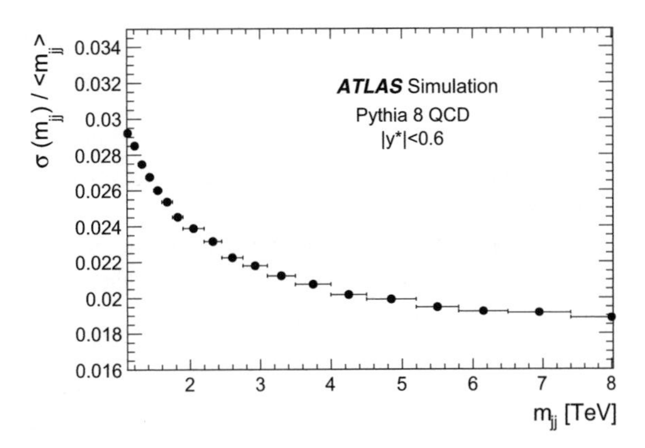

Fig. 5.17 The fractional dijet mass resolution $\frac{\sigma(m_{jj})}{\langle m_{jj}\rangle}$ versus m_{jj}^{truth}, taken from [39]

Fig. 5.17 we can see that the fractional dijet mass resolution starts to plateau at high mass. The extrapolation enables us to generate bins up to the centre-of-mass energy (13 TeV), allowing us to accommodate potential high mass signals, and ensuring that no events are missed. The extrapolated curve is used in the iterative procedure described below in order to derive the binning.

The starting point for deriving the binning (m_{initial}) was chosen to be 946 GeV. This value was chosen as it is above the m_{jj} peak in the QCD dijet MC with the looser p_T cut, and it is well below the 1.1 TeV starting point of the full analysis.

The iterative process proceeds as follows:

1. A guess at the position of the bin centre is made: $m_{\text{centre}} = m_{\text{initial}}(1 + \frac{1}{2}\frac{\sigma(m_{\text{initial}})}{\langle m_{\text{initial}}\rangle})$, where $\frac{\sigma(m_{\text{initial}})}{\langle m_{\text{initial}}\rangle}$ is the fractional dijet mass resolution evaluated at m_{initial}.

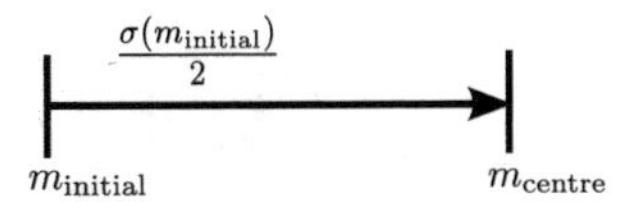

Note that the black arrow in the example diagram points from the starting point for the current step to the position calculated in this step.

2. The derived m_{centre} is used to calculate the lower bin edge: $m_{\text{lower}} = m_{\text{centre}}(1 - \frac{1}{2}\frac{\sigma(m_{\text{centre}})}{\langle m_{\text{centre}}\rangle})$ now using the fractional dijet mass resolution evaluated at m_{centre}.

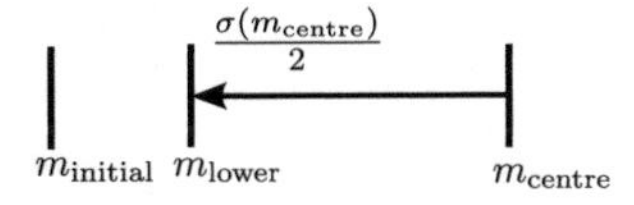

3. Check if m_{initial} (true bin edge) and m_{lower} (derived bin edge) agree to within a tolerance of 0.1%.
If $m_{\text{lower}} > m_{\text{initial}}$ shift m_{centre} by -0.01 GeV and repeat step 2.
If $m_{\text{lower}} < m_{\text{initial}}$ shift m_{centre} by $+0.01$ GeV and repeat step 2.

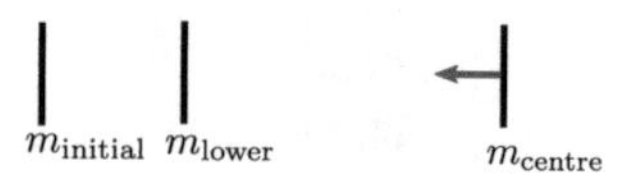

Note that the red arrow indicates the direction of the 0.01 GeV shift in the example diagram.

4. Once the tolerance threshold has been satisfied, the upper bin edge is calculated: $m_{\text{upper}} = m_{\text{centre}}(1 + \frac{1}{2}\frac{\sigma(m_{\text{centre}})}{\langle m_{\text{centre}}\rangle})$, and rounded to the nearest 1 GeV.

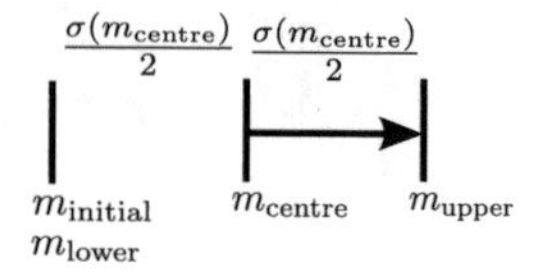

5. The procedure is repeated with $m_{\text{initial}} = m_{\text{upper}}$.

The final derived binning is given in Appendix B. The four initial criteria that the binning was required to satisfy were checked. The bin widths agreed with the resolution curve to within 1%, and the number of bins were maximised, satisfying criteria one and two. The bin widths were compared to the signal widths and were several times smaller, satisfying criteria 3. The final QCD dijet MC spectrum was smooth, satisfying criteria 4.

The Final Mass Spectrum from Data

The final mass spectrum for the high mass dijet analysis is shown in Fig. 5.18.

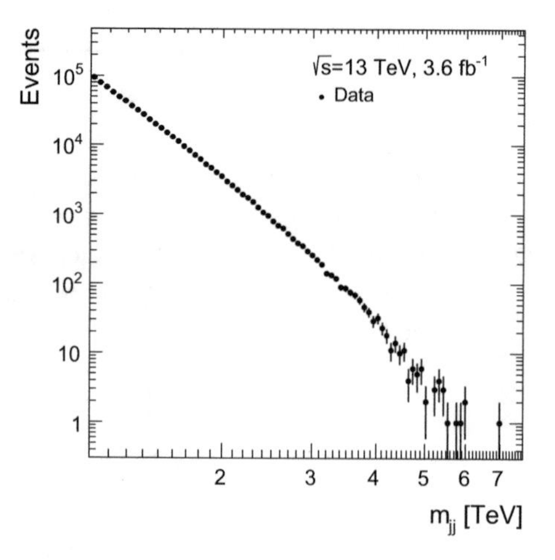

Fig. 5.18 The final mass spectrum for the high mass dijet analysis, using 3.6 fb^{-1} of $\sqrt{s} = 13$ TeV data

5.4.2 Spectra for the Dijet + ISR Analyses

In order to maintain consistency between both the dijet + ISR analyses, and the high mass dijet analysis, a common binning was used. In both of the dijet + ISR analyses the binning derived for the high mass dijet analysis is used, with an extension to low mass. In order to derive the extended binning in the low mass region, the same techniques, MC and analysis selections used to derive the high mass binning were applied, with the exception of the leading jet p_T cut. The leading jet p_T cut was lowered from 350 to 100 GeV in order to shift the m_{jj} peak in the QCD dijet MC lower in mass. The final derived binning is given in Appendix B.

A comparison was made between the fractional dijet mass resolution for each of the dijet + ISR analyses, and the fractional dijet mass resolution used to derive the extended binning. The comparison shows that the differences between the curves for

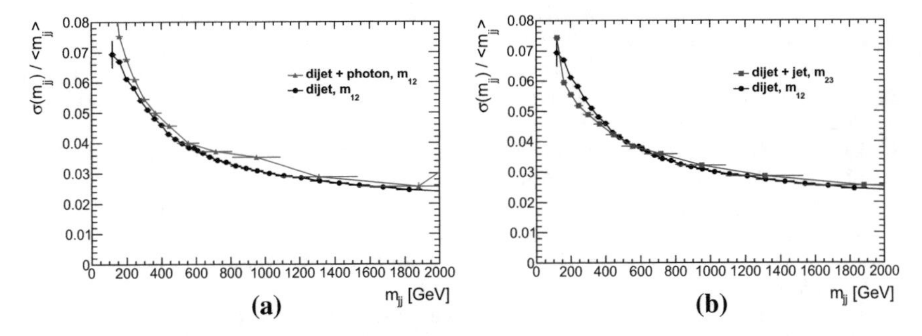

Fig. 5.19 The fractional dijet mass resolution $\frac{\sigma(m_{jj})}{\langle m_{jj}\rangle}$ versus m_{jj}^{truth} is shown. A comparison is made between the curve used to derive the extended binning, points shown in black and **a** the fractional dijet mass (m_{12}) resolution curve for the dijet + γ analysis shown in red, **b** the fractional dijet mass (m_{23}) resolution curve for the dijet + jet analysis shown in blue. Figure adapted from [40]

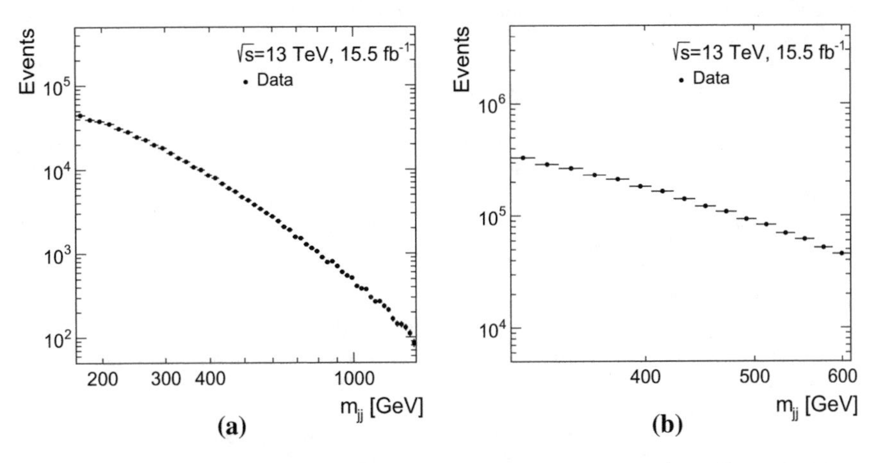

Fig. 5.20 The final mass spectra for the dijet + ISR analyses, using 15.5 fb^{-1} of $\sqrt{s} = 13$ TeV data, and triggered using an ISR **a** photon, **b** jet

the dijet + ISR analyses and for the curve used to derive the extended binning are small. Hence, it was decided that the extended dijet binning can be used for the dijet + ISR analyses (Fig. 5.19).

The Final Mass Spectra from Data

The final mass spectra for the dijet + ISR analyses are shown in Fig. 5.20.

References

1. ATLAS Collaboration (2016) Search for new phenomena in dijet mass and angular distributions from pp collisions at $\sqrt{s} = 13$ TeV with the ATLAS detector. Phys Lett B 754:302–322. https://doi.org/10.1016/j.physletb.2016.01.032, arXiv:1512.01530 [hep-ex]
2. Garvey J et al (2004) The ATLAS level-1 calorimeter trigger architecture. IEEE Trans Nucl Sci 51:356–360. https://doi.org/10.1109/TNS.2004.828800
3. Akesson T et al (2015) Search for new phenomena in dijet events with the ATLAS detector at $\sqrt{s} = 13$ TeV. Technical report, ATL-COM-PHYS-2015-290. Geneva: CERN
4. Sjostrand T, Mrenna S, Skands PZ (2008) A brief introduction to PYTHIA 8.1. Comput Phys Commun 178:852-867. https://doi.org/10.1016/j.cpc.2008.01.036, arXiv:0710.3820 [hep-ph]
5. ATLAS Collaboration (2014) ATLAS Run 1 Pythia8 tunes. Technical report, ATL-PHYS-PUB-2014-021. Geneva: CERN
6. Lange DJ (2001) The EvtGen particle decay simulation package. Nucl Instrum Meth A462:152–155. https://doi.org/10.1016/S0168-9002(01)00089-4
7. Ball RD et al (2013) Parton distributions with LHC data. Nucl Phys B867:244–289. https://doi.org/10.1016/j.nuclphysb.2012.10.003, arXiv:1207.1303 [hep-ph]
8. Particle Data Group, Patrignani C et al (2016) Review of particle physics. Chin Phys C40.10. https://doi.org/10.1088/1674-1137/40/10/100001
9. Alwall J et al (2014) The automated computation of tree-level and next-to-leading order diffierential cross sections, and their matching to parton shower simulations. JHEP 07:079. https://doi.org/10.1007/JHEP07(2014)079, arXiv:1405.0301 [hep-ph]
10. Abercrombie D et al (2015) In: Boveia A et al (ed) Dark matter benchmark models for early LHC Run-2 searches: report of the ATLAS/CMS dark matter forum. arXiv:1507.00966 [hep-ex]
11. Gallas E (1999) Tevatron searches for compositeness. Springer, Berlin, pp 1241–1244. https://doi.org/10.1007/978-3-642-59982-8_237
12. Dai D-C et al (2008) BlackMax: a black-hole event generator with rotation, recoil, split branes, and brane tension. Phys Rev D77:076007. https://doi.org/10.1103/PhysRevD.77.076007, arXiv:0711.3012 [hep-ph]
13. Gingrich DM (2010) Monte Carlo event generator for black hole production and decay in proton-proton collisions. Comput Phys Commun 181:1917–1924. https://doi.org/10.1016/j.cpc.2010.07.027, arXiv:0911.5370 [hep-ph]
14. Ashkenazi A et al (2013) ATLAS search in 2012 data for new phenomena in dijet mass distributions using pp collisions at $\sqrt{s} = 8$ TeV. Technical report, ATL-COM-PHYS-2013-1518. Geneva: CERN
15. Gingrich DM, Saraswat K (2012) Model uncertainties on limits for quantum black hole production in dijet events from ATLAS. arXiv:1210.3430 [hep-ph]
16. Pumplin J et al (2002) New generation of parton distributions with uncertainties from global QCD analysis. JHEP 07:012. https://doi.org/10.1088/1126-6708/2002/07/012, arXiv:hep-ph/0201195 [hep-ph]
17. Baak M et al (2010) Data quality status flags and good run lists for physics analysis in ATLAS. Technical report, ATL-COM-GEN-2009-015. Geneva: CERN

18. ATLAS Collaboration (2017) Data preparation check list for physics analysis. https://twiki.cern.ch/twiki/bin/view/Atlas/DataPreparationCheckListForPhysicsAnalysis
19. ATLAS Collaboration (2015) Selection of jets produced in 13 TeV proton-proton collisions with the ATLAS detector. Technical report, ATLAS-CONF-2015-029. Geneva: CERN
20. ATLAS Collaboration (2013) Characterisation and mitigation of beam-induced backgrounds observed in the ATLAS detector during the 2011 proton-proton run. J Instrum 8.07:P07004. https://doi.org/10.1088/1748-0221/8/07/P07004, arXiv:1303.0223 [hep-ex]
21. ATLAS Collaboration (2013) Jet energy measurement with the ATLAS detector in proton-proton collisions at $\sqrt{s} = 7$ TeV. Eur Phys J C73.3:2304. https://doi.org/10.1140/epjc/s10052-013-2304-2, arXiv:1112.6426 [hep-ex]
22. High mass dijet analysis team (2015), Private communication
23. An H, Huo R, Wang L-T (2013) Searching for low mass dark portal at the LHC. Phys Dark Univ 2:50–57. https://doi.org/10.1016/j.dark.2013.03.002, arXiv:1212.2221 [hep-ph]
24. Shimmin C, Whiteson D (2016) Boosting low-mass hadronic resonances. Phys Rev D94.5:055001. https://doi.org/10.1103/PhysRevD.94.055001, arXiv:1602.07727 [hep-ph]
25. ATLAS Collaboration (2016) Search for new light resonances decaying to jet pairs and produced in association with a photon in proton-proton collisions at $\sqrt{s} = 13$ TeV with the ATLAS detector. Technical report, ATLAS-CONF-2016-029. Geneva: CERN
26. ATLAS Collaboration (2015) Search for new light resonances decaying to jet pairs and produced in association with a photon or a jet in proton-proton collisions at $\sqrt{s} = 13$ TeV with the ATLAS detector. Technical report, ATLAS-CONF-2016-070. Geneva: CERN
27. Gleisberg T et al (2009) Event generation with SHERPA 1.1. JHEP 02:007. https://doi.org/10.1088/1126-6708/2009/02/007, arXiv:0811.4622 [hep-ph]
28. Höche S et al (2009) QCD matrix elements and truncated showers. JHEP 05:053. https://doi.org/10.1088/1126-6708/2009/05/053, arXiv:0903.1219 [hep-ph]
29. Pumplin J et al (2010) New parton distributions for collider physics. Phys Rev D82:074024. https://doi.org/10.1103/PhysRevD.82.074024, arXiv:1007.2241 [hep-ph]
30. Buttinger W (2015) Using event weights to account for difierences in Instantaneous Luminosity and trigger prescale in Monte Carlo and data. Technical report, ATL-COM-SOFT-2015-119. Geneva: CERN
31. Mistry K, Thompson AJ, Williams H (2016) Data-MC shower shape comparisons: supporting documentation for the photon identification in 2015 ATLAS data. Technical report, ATL-COM-PHYS-2016-574. Geneva: CERN
32. ATLAS Collaboration (2014) Tagging and suppression of pileup jets with the ATLAS detector. Technical report, ATLAS-CONF-2014-018. Geneva: CERN
33. ATLAS Collaboration (2016) Electron and photon energy calibration with the ATLAS detector using data collected in 2015 at $\sqrt{s} = 13$ TeV. Technical report, ATLAS-PHYS-PUB-2016-015.Geneva: CERN
34. ATLAS Collaboration (2014) Electron and photon energy calibration with the ATLAS detector using LHC Run 1 data. Eur Phys J C 74.10:3071. https://doi.org/10.1140/epjc/s10052-014-3071-4, ISSN: 1434-6052
35. ATLAS Collaboration (2016) Photon identification in 2015 ATLAS data. Technical report, ATLPHYS- PUB-2016-014. Geneva: CERN
36. ATLAS Collaboration Isolation Forum (2016) Official isolation working points. https://twiki.cern.ch/twiki/bin/view/AtlasProtected/IsolationSelectionTool#Photons
37. ATLAS Collaboration (2015) Recommendations of the physics objects and analysis harmonisation study groups 2014. Technical report, ATL-COM-PHYS-2014-451. Geneva: CERN
38. Callea G, Pitt M (2016) Photon identification efficiency measurement with the matrix method: supporting documentation for the Photon identification in 2015 ATLAS data. Technical report, ATL-COM-PHYS-2016-573. Geneva: CERN
39. ATLAS Collaboration (2016) Auxiliary material for search for new phenomena in dijet mass and angular distributions from pp collisions at $\sqrt{s} = 13$ TeV with the ATLAS detector. https://atlas.web.cern.ch/Atlas/GROUPS/PHYSICS/PAPERS/EXOT-2015-02/
40. Corrigan E (2017) Private communication

Chapter 6
Searching for Resonances

The common goal of the high mass dijet analysis and the dijet + ISR analyses is to determine if there is evidence for resonances being present in the spectra, and to assess their significance. In order to determine if such resonances are present it is necessary to compare the spectra produced in Chap. 5 to an estimate of the background produced via Standard Model processes. The background estimations utilised in these analyses are described in Sect. 6.2. Once the background estimates have been derived, a comparison between the background estimate and the data needs to be made for each analysis, and a quantitative measure of the significance of the largest excess is calculated, as described in Sect. 6.3. Validations of the methods are given in Sect. 6.4, before the final results are presented in Sect. 6.5. Before proceeding, a brief overview of the statistical approach utilised in this *search phase* of the analyses will be given in Sect. 6.1.

6.1 Statistical Approach

There are two main notions of probability utilised in particle physics: a frequentist approach, and a Bayesian approach. In the search phase the frequentist approach is utilised.

In the frequentist approach, the underlying parameters of a theory are considered to be fixed and a large number of pseudo-experiments are conducted [1]. In our case, the background estimate represents our underlying theory, and we generate pseudo-experiments as follows: for each bin in the background fit, a draw is made from a Poisson distribution with the parameter of the distribution set to the bin content of the fit. A pseudo-experiment is a full set of these single bin draws, producing a new spectrum. Each pseudo-experiment represents an outcome which could be obtained if our theory is true. By comparing the outcome obtained using data to the relative frequency of the outcome of the pseudo-experiments, the probability of

L. A. Beresford, *Searches for Dijet Resonances*, Springer Theses,
https://doi.org/10.1007/978-3-319-97520-7_6

obtaining the data under the assumption of the theory can be calculated [1]. In the search phase this approach is utilised to obtain the probability that an excess with greater or equal significance than what we observe in our data could be produced under the background only hypothesis. In order to calculate the probability, referred to as a p-value, a test statistic is used. A test statistic is a function of the data used to quantify the level of agreement between the data and a hypothesis; often it is a single value which increases with increasing discrepancy [2]. In our case the test statistic is used to identify the region of the dijet invariant mass spectrum with the largest excess, and to indicate the size of the excess. The value of the test statistic in data is compared to the distribution of test statistics obtained using pseudo-experiments generated from the background estimate. This allows us to calculate the fraction of pseudo-experiments with larger test statistics than we observed in data, and hence to calculate the p-value. A small p-value indicates a deviation from the background only hypothesis. Further details about this procedure will be given in this chapter. A description of the Bayesian approach will be given in Chap. 7.

6.2 Background Estimate

The background distributions for each of the dijet analyses is the result of the underlying physics processes (mainly QCD), the quark and gluon PDFs, detector resolution effects and analysis selections. A method for accurately describing this background distribution is vital to perform the analyses.

Many analyses use Monte Carlo to estimate the background shape; however, in the dijet analyses considered here this is not practical. Due to the high number of events in the spectra, a very accurate description of the background is needed, and MC may not model the shape of the background contribution with sufficient accuracy, due to the difficulty of modelling QCD processes. Additionally, there are substantial theoretical uncertainties associated with the use of QCD MC, which would reduce the sensitivity of the analysis. Finally, due to the high QCD cross-section, a huge number of Monte Carlo events would need to be generated in order to obtain sufficient statistical precision, which is both costly and time-consuming. Due to the reasons given, many of the dijet resonance searches performed over the last 27 years have employed the data-driven approach of using a smooth fit to parametrise the background [3].

The use of a fit function has been shown to be successful in previous dijet analyses [3]; however, it is crucial that the fit function is chosen with care. Important considerations include: the form of the function used, the number of free parameters, and the range in mass over which the fit spans.

The chosen function must be a smoothly falling monotonic function which is capable of describing the background shape and the high mass tail, in the case of the high mass dijet analysis. Historically, functions resembling the following form have been used to describe dijet invariant mass spectra:

$$f(x) = p_1(1-x)^{p_2}x^{p_3+p_4\ln(x)+p_5(\ln(x))^2}, \tag{6.1}$$

where $x \equiv \frac{m_{jj}}{\sqrt{s}}$ and p_i are the fit parameters. Note that the fit parameters can be set to zero in order to decrease the degrees of freedom of the fit, if they are not needed. The functional form was originally motivated by the leading order QCD matrix elements, with the introduction of a dependence on $m_{jj}^{p_i}$ in the functional form used by the UA2 Collaboration in 1991 [3, 4]. The function was further developed through the introduction of a dependence on $(1-\frac{m_{jj}}{\sqrt{s}})^{p_i}$ by the CDF Collaboration in 1995, in order to model the mass dependent behaviour of parton distribution functions [3, 5]. Since then the functional form has evolved based on empirical observations, in order to fit more precise data which extends to higher mass regions. This functional form is utilised in the analyses described in this thesis, and the fit range and number of parameters are selected in a data-driven way. Before discussing the approaches used to determine the fit range and the fit function, i.e. the number of free parameters to be used in Eq. (6.1), a description of the impact of the blinding strategy on the background estimation strategy will be given, followed by a brief description of the fitting implementation.

6.2.1 Blinding Strategy Impact on the Background Estimation

The un-blinded approach adopted by the analyses in this thesis is beneficial for the background estimation strategy. By analysing datasets of increasing sizes and fixing our background estimation procedure prior to data taking, we were able to automatically change the number of parameters in the background parametrisation, if needed, as the dataset size increased. In previous dijet analyses, it has been observed that more statistically precise data may require more parameters in the fit in order to accurately describe the background. One such analysis was the dijet resonance search performed by the ATLAS Collaboration using $\sqrt{s} = 8$ TeV data [6]. In this analysis, the background parametrisation and the corresponding number of parameters was selected in advance. The procedure was tested on the 'blind' dataset, i.e. one quarter of the total dataset in this case. The chosen parametrisation could accurately describe the blind data set; however, when used to describe the full dataset, it was seen that the chosen parametrisation provided a poor description of the background. It was also seen that by including an additional parameter, a much better description could have been obtained. By allowing our parametrisation to 'evolve' with the dataset luminosity this problem can be avoided. Additionally, for the dijet + ISR analyses, which aim to search as low in mass as possible, a data-driven approach was used to determine the starting position for the fit.

6.2.2 Fitting Implementation

In both the high mass dijet analysis and the dijet + ISR analyses, a Maximum Likelihood Estimator (MLE) fit [7, 8] was utilised. This type of fit was utilised as it can be applied in regions where there are very few events, e.g. the tail of the high mass spectrum. Other fit methods, such as the χ^2 fit [9], should not be applied in regions where there are fewer than ∼5 events [10].

The likelihood function used is denoted by $\mathcal{L}(\nu|\text{Data})$, where ν is the set of fit function parameters. This is equivalent to the probability of obtaining our spectrum, given the set of parameters, i.e. $P(\text{Data}|\nu)$ [2]. Since the data in each bin of the dijet spectra are Poisson-distributed, the likelihood function is defined as the product of the Poisson probability in each bin:

$$\mathcal{L}(\nu|\text{Data}) = \prod_{\mathrm{i}=1}^{\mathrm{N}} \frac{\mathrm{B_i^{D_i}e^{-B_i}}}{\mathrm{D_i!}}, \tag{6.2}$$

where the product runs over all the bins i in the spectrum, B_i is the number of background events in bin i, calculated through integration of the fit function with parameters ν. D_i is the number of data events in bin i. By maximising $\mathcal{L}(\nu|\text{Data})$, or equivalently minimising $-\ln(\mathcal{L}(\nu|\text{Data}))$ [1], we obtain the best estimate of ν. The minimisation approach was utilised, as it is more convenient when the likelihood needs to be computed over many bins. MINUIT [11] was used to perform the minimisation.

6.2.3 Selection of Fit Range

A key factor to consider is the range over which the function is applied, as this determines the region of the mass spectrum in which we can derive a background estimate, and hence, the region in which we can search. As stated previously, for the high mass dijet analysis there is a requirement that the dijet invariant mass is 1.1 TeV or above, to avoid any bias from the trigger efficiency. Therefore, the chosen fit range for this search is from 1.1 TeV to the highest dijet mass observed in the spectrum.

For the dijet + ISR analyses, the aim is to search as low in mass as possible. The lower bound is dictated by where the dijet mass spectrum becomes sculpted by the kinematic selections. In order to determine how low in mass we can fit, we first need to fix the upper bound of the fit range. For the dijet + γ analysis, an upper bound of 1493 GeV was chosen, as this allows some overlap with the mass range probed by the high mass dijet analysis. For the dijet + jet analysis, the choice of end point is a bit more complicated. As mentioned in Sect. 5.2.2, at higher masses m_{23} becomes the 'wrong combination' for reconstructing the resonance. This results in the formation of 'tails' around the resonance peak, as illustrated in Fig. 6.1. Therefore, mass points above 550 GeV are not considered, as a sharp signal peak is no longer produced. An

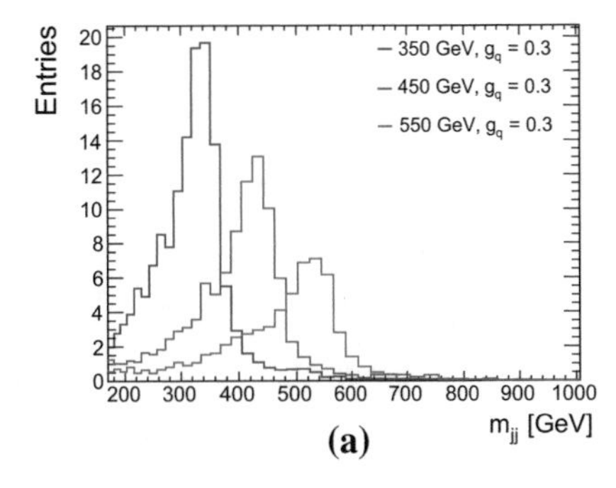

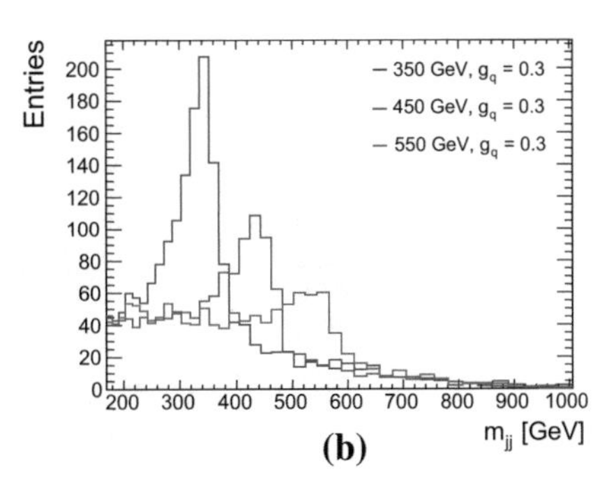

Fig. 6.1 A signal shape comparison for three mass points, 350, 450 and 550 GeV for the Z' model with $g_q = 0.3$, for **a** the dijet + γ analysis; and **b** the dijet + jet analysis. It is seen that for the dijet + jet analysis large tails are produced for higher mass points due to combinatoric effects; however, this is not the case for the dijet + γ analysis

upper bound of 611 GeV was chosen for the fit range. Note that the values 1493 and 611 GeV correspond to bin edges.

Now that the upper bound of the fit range has been determined, the lower bound can be chosen by incrementally decreasing the lower bound in steps of 10 GeV and calculating the χ^2 p-value for each fit. The χ^2 value is used as the test statistic. It quantifies differences between the data and the fit and increases with increasing discrepancy. The probability of obtaining a test statistic which is equal to or higher than our observed test statistic, under the assumption that the null hypothesis is true, i.e. that the m_{jj} spectrum is described by the fit function, is called the p-value. The p-value is calculated as the fraction of the χ^2 distribution which has a χ^2 value which is equal to or higher than our observed value. A small p-value indicates a poor fit to the data.

Figure 6.2 shows the χ^2 p-value versus the lower bound of the three, four and five parameter fit functions applied to the dijet + ISR spectra. A lower bound of 169 GeV was selected for the dijet + γ analysis and a lower bound of 303 was selected for the dijet + jet analysis. These values correspond to bin edges.

6.2.4 Selection of Fit Function

As previously mentioned, since the background estimate is determined by the fit function, it is absolutely critical that an appropriate function is chosen. If a function with too few parameters is used then an accurate estimate of the background will not be obtained, and disagreements between the data and the background estimate could produce *spurious signals*. Alternatively, if a function with too many parameters is used then the function may be too flexible and signals may become incorporated in the background estimate.

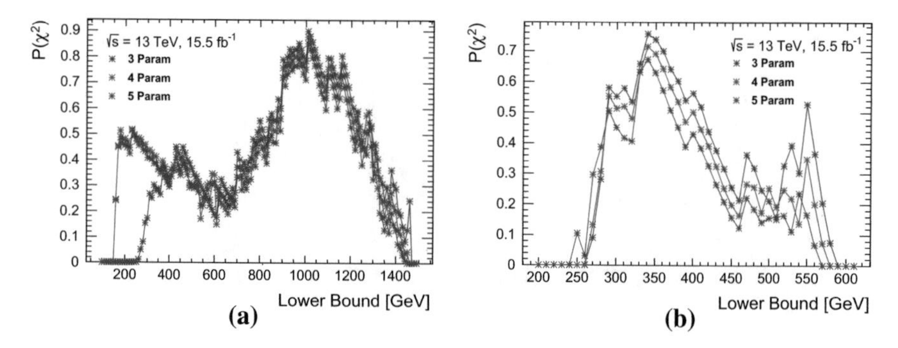

Fig. 6.2 The χ^2 p-value for the three, four and five parameter fits, as a function of the lower bound of the fit for **a** the dijet + γ analysis; and **b** the dijet + jet analysis. For the dijet + γ anlysis the upper bound of the fit is fixed to 1493 GeV and for the dijet + jet analysis the upper bound of the fit is fixed to 611 GeV

The chosen functional form is given in Eq. (6.1), and was selected as functions of this form have been shown to provide satisfactory descriptions of dijet invariant mass spectra from many dijet analyses since the 1990s (with lower centre-of-mass energies) [3]. Additionally, studies in Monte Carlo showed that functions of this form were able to describe $\sqrt{s} = 13$ TeV dijet invariant mass spectra produced using Monte Carlo. However, there is no guarantee that it will be able to provide a satisfactory description of the $\sqrt{s} = 13$ TeV dijet invariant mass spectra from data shown in Chap. 5.

As previously mentioned, a robust strategy was developed in order to allow the fit function choice to evolve with luminosity. The strategy uses the Wilks' test [12].

The Wilks' Test

The Wilks' test is used to determine if an alternate hypothesis H_1 provides a significantly better description of our data than the null hypothesis H_0. In our case, H_0 is the background estimate obtained using a fit with n degrees of freedom, and H_1 is the background estimate obtained using $n+1$ degrees of freedom. For example, if H_0 is obtained using the three parameter fit function (setting p_4 and p_5 to zero in Eq. (6.1)), then H_1 is obtained using the four parameter fit function (setting p_5 to zero in Eq. (6.1)). A comparison between the hypotheses is made by employing a test statistic based on the likelihood ratio Λ of the two hypothesis. The test statistic is defined as:

$$-2\ln(\Lambda) = -2\ln\left(\frac{\mathcal{L}(\mathrm{H_0}|\mathrm{Data})}{\mathcal{L}(\mathrm{H_1}|\mathrm{Data})}\right). \tag{6.3}$$

This is a single number which increases as the maximum likelihood of H_1 increases with respect to the maximum likelihood of H_0. In order to quantify the significance of the obtained value of the test statistic, we need to compare it to the predicted distribution under the assumption that H_0 is true. The predicted distribution is obtained using Wilks' Theorem [12]:

Wilks' Theorem: For nested models, under the assumption that H_0 is true, as the sample size approaches ∞, the test statistic $-2\ln(\Lambda)$ approaches a χ^2 distribution with number of degrees of freedom equal to the difference in the number of parameters of the two models.

The p-value can then be obtained in the same way as before, i.e. by calculating the fraction of the predicted distribution which has a value of the test statistic equal to or greater than the value we observe. The p-value is then used to decide whether to reject H_0 in favour of H_1, as a small p-value indicates that H_1 describes the data better than H_0.

The Full Procedure

For each increase in dataset size, the choice of fit function proceeds as follows:

1. Before performing the Wilks' test, the data is checked for significant deviations. The baseline 3 parameter function is fit to the data, and the BumpHunter p-value is calculated. If the calculated p-value is < 0.01, indicating there is a local excess, then a window spanning the excess is removed. The BumpHunter p-value and window removal procedure are described in Sect. 6.3.2. The Wilks' test is then performed using the remaining bins.
2. If the p-value from the Wilks' test is < 0.05, indicating an unsatisfactory fit, the next highest order function is chosen as the baseline and the procedure is repeated until a satisfactory fit function is chosen.

The evolution of the Wilks' p-value with increasing luminosity is shown in Fig. 6.3 for the high mass dijet analysis. The blue curve shows the Wilks' p-value calculated with the 3 parameter fit function as the null hypothesis and the 4 parameter fit function as the alternative hypothesis. Since the blue curve stays above the 0.05 threshold, indicated by the dashed line, for all dataset sizes, and the final Wilks' p-value for the full dataset is 0.77. This shows that the 3 parameter function is sufficient for this analysis.

The same procedure is used for the dijet + ISR analyses, and the corresponding Wilks' p-value plots are shown in Fig. 6.4. From these plots it is seen that the 4 parameter function is sufficient for the dijet + γ analysis, and the 3 parameter function is sufficient for the dijet + jet analysis.

6.3 Assessing Significance

In addition to deriving a background estimate for each analysis, a method for comparing the obtained background estimate and the data is needed in order to search for excesses in the spectra. The significance of the largest excess observed in the spectra must then be quantified. The test statistic must be chosen with care, such that it is appropriate for our needs. The χ^2 test statistic, which we introduced earlier, is unsuitable to find the exact location of the largest excess and to quantify its significance. This test statistic utilises bin-by-bin quantities in order to quantify the level of

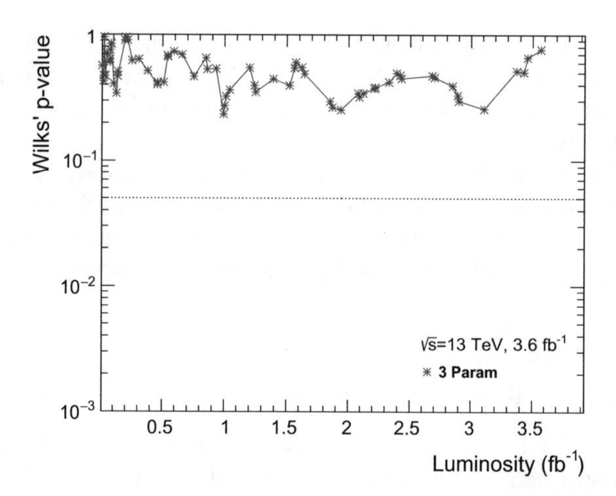

Fig. 6.3 The evolution of the Wilks' p-value with luminosity for the full 3.6 fb^{-1} dataset used in the high mass dijet analysis. The blue curve compares the 3 parameter fit function to the alternate 4 parameter function, showing that the 3 parameter function is sufficient. This figure is adapted from [13]

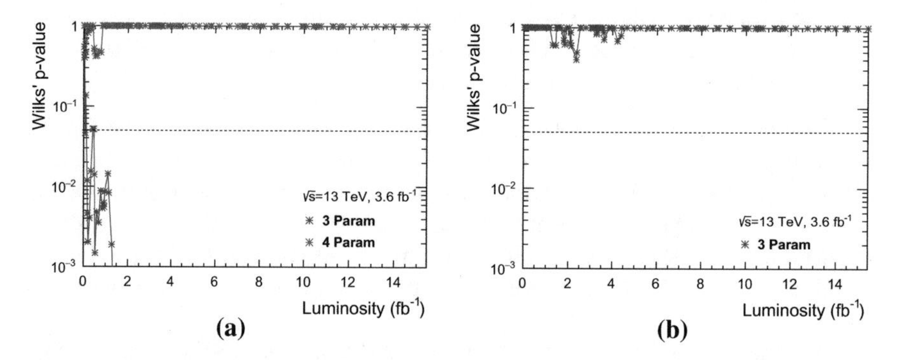

Fig. 6.4 The evolution of the Wilks' p-value with luminosity for the full 15.5 fb^{-1} dataset, for **a** the dijet + γ analysis, and **b** the dijet + jet analysis. The blue curve compares the 3 parameter fit function to the alternate 4 parameter function, and the red curve compares the 4 parameter fit function to the alternate 5 parameter function

agreement, without taking into account the relative positions of the excesses. As we are interested in searching for excesses across neighbouring bins, we would like to utilise a test statistic which takes into account the position of excesses with respect to one another.

A suitable test statistic to use is provided by the BumpHunter algorithm [14], described in Sect. 6.3.1. The full search procedure is described in Sect. 6.3.2, and a description of how we can visually display the excesses and deficits in the spectra is given in Sect. 6.3.3.

6.3.1 *The* BUMPHUNTER *Algorithm*

The BUMPHUNTER algorithm scans the distribution looking for excesses and deficits in all sets of neighbouring bins, referred to as windows. Window widths ranging from two bins wide to windows utilising half the bins in the spectrum are considered in the search. For each window, a local p-value is calculated using the Poisson probability of obtaining an excess or deficit larger than the observed excess or deficit. The local p-value is defined as

$$\text{p-value} = \begin{cases} \sum\limits_{\mathrm{n=D}}^{\infty} \frac{\mathrm{B}^{\mathrm{n}} e^{-\mathrm{B}}}{\mathrm{n}!}, & \text{for D} \geq \text{B} \\ \sum\limits_{\mathrm{n=0}}^{D} \frac{\mathrm{B}^{\mathrm{n}} e^{-\mathrm{B}}}{\mathrm{n}!}, & \text{for D} < \text{B} \end{cases} \tag{6.4}$$

where D is the total number of data events in the window, and B is the total number of background events in the window. This probability can be calculated analytically, without the need for pseudo-experiments, making the calculation of the local p-value quick, as described in [14].

An example plot showing the local p-value for each window is shown in Fig. 6.5. The plot was produced by applying BUMPHUNTER to a dijet mass distribution generated using PYTHIA truth level dijet Monte Carlo with no BSM signal added. The MC distribution used here is *data-like*, i.e. it has an equivalent statistical accuracy as data with the same luminosity. A description of the production of data-like spectra will be given in Sect. 6.4.

The test statistic t utilised by the BUMPHUNTER algorithm is defined as

$$t = -\log(\text{p-value}_{\text{min}}), \tag{6.5}$$

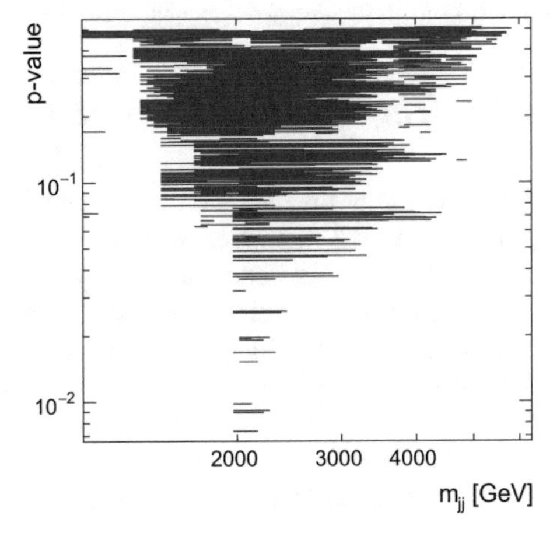

Fig. 6.5 Each line indicates a window considered in the search. The horizontal position of the line and the length of the line indicate the position and width of the window in mass, respectively. The vertical position of the line indicates the local p-value for the window. In order to obtain this plot the BUMPHUNTER algorithm was applied to a data-like dijet mass spectrum generated using Monte Carlo

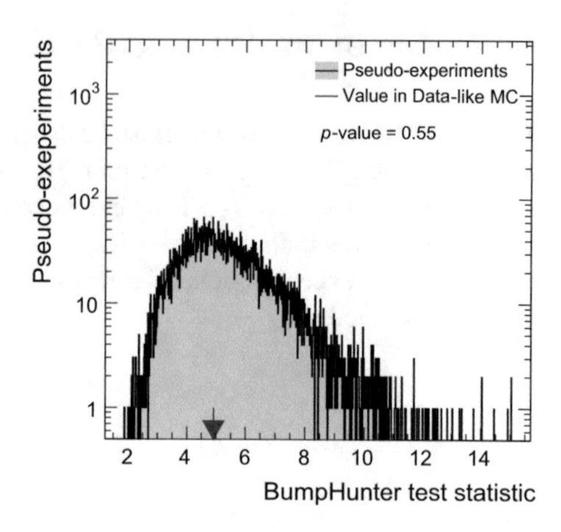

Fig. 6.6 The distribution of BUMPHUNTER test statistics from pseudo-experiments is shown, and the observed value of the BUMPHUNTER test statistic in data-like MC is indicated by the arrow. The global p-value is derived by calculating the fraction of pseudo-experiments with a BUMPHUNTER test statistic greater than the observed value of the BUMPHUNTER test statistic

where p-value$_{min}$ is the smallest local p-value for any window considered in the search. This test statistic is used to calculate the global p-value for the most discrepant window in the spectrum. In contrast to the local p-value, which considers the probability of an excess arising at a specific position in the spectrum, the global p-value accounts for the fact that we are searching for an excess at any position in the mass spectrum, increasing the probability of observing an excess; this is the basic idea behind the *look-elsewhere effect* [2].

The global p-value is used to decide whether to reject the background only hypothesis, rather than the value of the local p-value. The global p-value is calculated by comparing the observed t to the distribution of t obtained using 10,000 pseudo-experiments generated from the background estimate, as illustrated in Fig. 6.6. The global p-value calculated using BUMPHUNTER accounts for the look-elsewhere effect since t, given in Eq. (6.5), considers every window in the search. Hence, we are calculating the probability that, in the background only case, there is an excess at any position in the spectrum at least as significant as the one observed in data [15].

6.3.2 The Full Search Procedure

A description of the method for choosing the fit range and parametrisation has been provided, as well as a description of the method for identifying excesses and quantifying their significance. Additionally, we must ensure that any significant excess in the spectrum does not distort the background estimate, which could affect the calculation of the significance of the excess. In these analyses, a significant excess is defined as having a p-value < 0.01, indicating that there is less than 1% chance this fluctuation could have occurred through statistical fluctuations of the background alone [16]. The full search procedure addresses the calculation of the background estimate in the presence of a significant signal, and proceeds as follows:

1. Fit the spectrum to obtain the background estimate.

2. Calculate BUMPHUNTER p-value using 10,000 pseudo-experiments.
3. If BUMPHUNTER p-value $\geq$ 0.01, procedure stops. There are no significant excesses present, proceed to limit setting.
4. If BUMPHUNTER p-value <0.01, a window is excluded in the calculation of the background estimate. Obtain initial exclusion window identified by BUMPHUNTER, i.e. the window with smallest local p-value. Note that if the window is at the start of the fit range then the window with the second smallest local p-value is used instead.
5. Re-fit the spectrum with the window masked in the fit. Calculate the BUMPHUNTER p-value for the remaining spectrum. If the BUMPHUNTER p-value for the remaining spectrum is still <0.01:
   - Obtain window with smallest local p-value in the remaining spectrum.
   - If this window is adjacent to the masked window, then mask one extra bin on the side of the remaining excess. Otherwise, mask one extra bin on both sides of the masked window.
   - Repeat this step until the BUMPHUNTER p-value $\geq$ 0.01, or the background estimate cannot be improved further, i.e. the masked region is > half the spectrum, or it is less than 2 bins away from the start or end of the spectrum.
6. One additional bin is masked at the low mass side of the masked region. This decision was made based on studies using MC signals, which showed that this improves the fit by removing any residual bias from the signal. The spectrum is re-fit with the final masked region.
7. The background estimate has now been computed and BUMPHUNTER is run on the full spectrum to obtain the final BUMPHUNTER p-value. Note that the BUMPHUNTER p-value does not take into account systematic uncertainties.

6.3.3 Bin-by-Bin Significances

In addition to quantifying the significance of the largest excess using the BUMPHUNTER p-value, it is useful to display the bin-by-bin significances of differences between the data and the background estimate. In order to display these differences, the significance in standard deviations is used, following the proposal described in detail in [17]. In order to calculate the significance of deviations, the Poisson p-value defined in Eq. (6.4) is used; however, D and B are calculated for each bin, rather than in a window.

As the p-value can span many orders of magnitude, it is convenient to use it to calculate a z-value. A z-value is the equivalent significance of the p-value, expressed as a standard deviation on the right hand side of a Gaussian distribution. The equation for the translation between the two is given by

$$\text{p-value} = \int_{\text{z-value}}^{\infty} \frac{1}{\sqrt{2\pi}} e^{-\frac{x^2}{2}} dx. \tag{6.6}$$

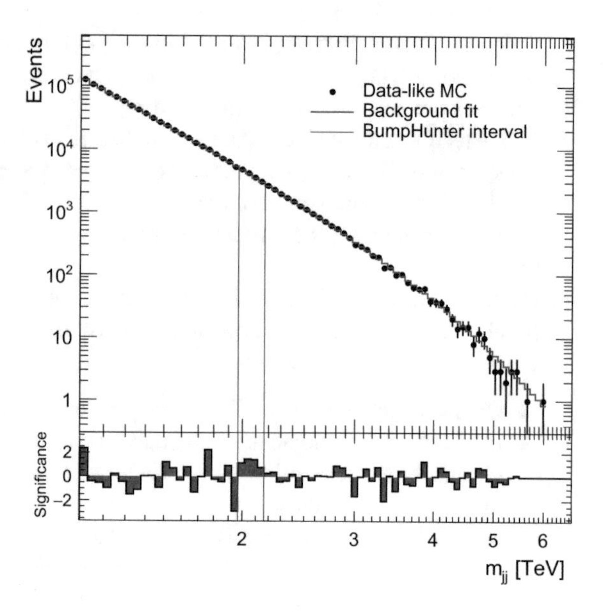

Fig. 6.7 In this plot the data-like MC dijet invariant mass spectrum is shown by the black points. The background estimate is shown by the red curve, and the blue vertical lines indicate the window with the largest significance. The lower panel shows the bin-by-bin significances of excesses and deficits in the data-like spectrum, measured in standard deviations

Only deviations with positive z-values, corresponding to p-values ≤ 0.5, are considered, as other deviations would be extremely minor and would complicate the procedure for displaying the significances, as explained in [17]. The displayed z-values are assigned a sign, whereby excesses in the data are displayed as positive z-values, and deficits in the data are displayed as negative z-values. Larger magnitude z-values, e.g. $\geq 3\sigma$, indicate a significant discrepancy between the data and the background estimate.

An example plot showing the results of the application of the full search procedure and the bin-by-bin significances panel is given in Fig. 6.7. In the top panel, the black points show the data-like m_{jj} distribution generated using Monte Carlo, the background estimate is shown by the red curve, and the blue vertical lines indicate the window with the largest significance found using BumpHunter. The lower panel shows the bin-by-bin significances for this distribution.

6.4 Method Validation

In order to validate the methods used to estimate the background and to perform the search, many tests were performed using Monte Carlo. In Sect. 6.4.1, we apply the search procedure to data-like MC spectra to ensure that we are able to accurately describe the background. In Sect. 6.4.2, the search procedure is applied to data-like MC background spectra combined with a signal to ensure that we are able to correctly identify significant signals, and that the background estimate is robust in the presence of signals. In Sect. 6.4.3, the test performed to check for the introduction of spurious signals is described. Note that in this section Monte Carlo scaled to the luminosity of the full dataset is used; however, prior to performing the search procedure on data, tests were conducted using smaller datasets.

6.4.1 Applying the Search Procedure to Monte Carlo

By using MC to produce a spectrum which closely resembles the spectrum obtained using data, referred to as a data-like spectrum, we can apply the search procedure to this spectrum to ensure that we obtain reliable results and are able to accurately describe the background.

Generating Data-Like Spectra

Monte Carlo events are generated such that an event level weight needs to be applied in order to produce distributions with a meaningful shape, the distributions are then weighted overall or in slices, in order to obtain the correct normalisation. However, the data in each bin of the m_{jj} spectrum contains events with unit weight and a statistical uncertainty equal to the square root of the total number of events in the bin. Our goal is to obtain spectra produced using MC with the same characteristics as data, i.e. data-like spectra which correctly replicate the statistical uncertainty expected when using data.

The method given in [18] is utilised to produce the spectra, and a description of this method is provided here. The basic idea is that we select a fraction of our total MC events, so we have a MC sample containing the same number of events as we expect in data for our chosen luminosity. We then assign unit weight to each of the events in our MC sample. A precise description of the method will now be given.

For each bin in the m_{jj} spectrum:

1. The condition $N_{\mathrm{effective}} > N_{\mathrm{weighted}}$ is checked, where N_{weighted} is the number of weighted MC events, and $N_{\mathrm{effective}}$ is the effective number of entries, given by $N_{\mathrm{effective}} = \frac{(\Sigma \mathrm{weights})^2}{\Sigma \mathrm{weight}^2} = \left(\frac{\mathrm{bin\ content}}{\mathrm{bin\ error}}\right)^2$. Note that if $N_{\mathrm{effective}}$ was calculated for data rather than MC, then $N_{\mathrm{effective}}$ would be equal to the number of data events, as the bin error would be equal to the square root of the bin content.
2. If the condition is satisfied then there is sufficient MC events to apply the data-like method and the ratio $r = \frac{N_{\mathrm{weighted}}}{N_{\mathrm{effective}}}$ is calculated.
3. For each MC event in the bin, a random number drawn from a flat distribution between 0 and 1 is assigned. If the random number is less than r then the event is kept and assigned unit weight, otherwise it is rejected.

Each bin in the m_{jj} spectrum has its own unique seed for the random number generator. By utilising the same bin-dependent seed when producing data-like spectra corresponding to different luminosities, the process of data collection is simulated, i.e. the events in lower luminosity spectra are a sub-set of the events in the higher luminosity spectra, like in data. This feature is important in tests of the fitting procedure; for example, the Wilks' test, which only makes sense if data is added to lower luminosity spectra to produce the higher luminosity spectra, rather than each spectrum being independent.

Results from Applying the Search Procedure to Monte Carlo

Data-like MC spectra generated using this method were used to test the search procedure for each analysis. For the high mass dijet analysis and the dijet + jet analysis the PYTHIA truth level MC sample was used to generate the spectra as it has significantly more events than the reconstructed level MC sample, allowing us to produce data-like spectra corresponding to the data luminosity. For the dijet + γ analysis, the reconstructed level SHERPA MC sample had sufficient events to produce a data-like spectrum corresponding to the data luminosity.

Examples of testing the search procedure for the high mass dijet analysis using a data-like spectrum corresponding to 3.6 fb^{-1} were shown in Figs. 6.5, 6.6 and 6.7. A global p-value of 0.55 was obtained, correctly indicating that there is no significant excess in the spectrum. The goodness of fit indicated by the χ^2 p-value was calculated to be 0.85, showing that the fit is of good quality. For the dijet + ISR analyses, the search procedure was applied to data-like MC spectra corresponding to 15.5 fb^{-1}, and the results are shown in Fig. 6.8.

For the dijet + γ analysis, a global p-value of 0.89 was obtained, and for the dijet + jet analysis, a global p-value of 0.25 was obtained In both cases this indicates that no significant excess was observed, as expected. In both cases the quality of the fits were good, as indicated by a χ^2 p-value of 0.92 for the dijet + γ analysis and 0.34 for the dijet + jet analysis.

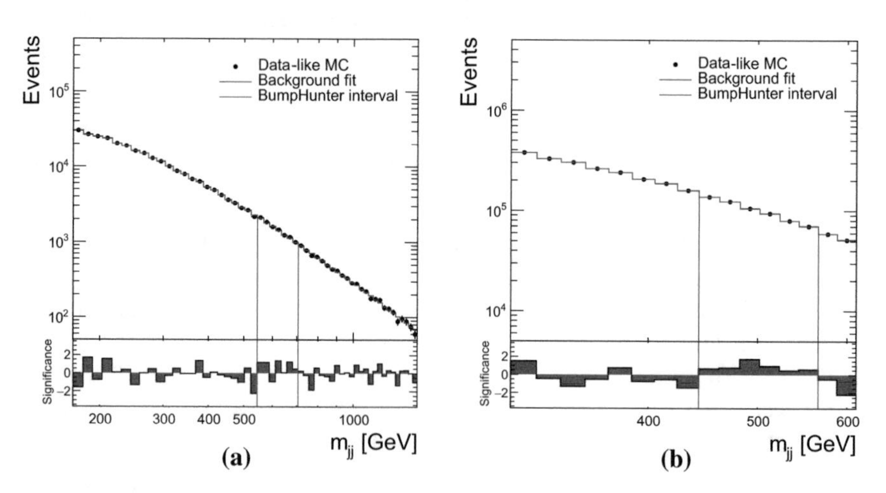

Fig. 6.8 These plots show the results of applying the search procedure to a data-like MC dijet invariant mass spectrum corresponding to 15.5 fb^{-1} for **a** the dijet + γ analysis, and **b** the dijet + jet analysis

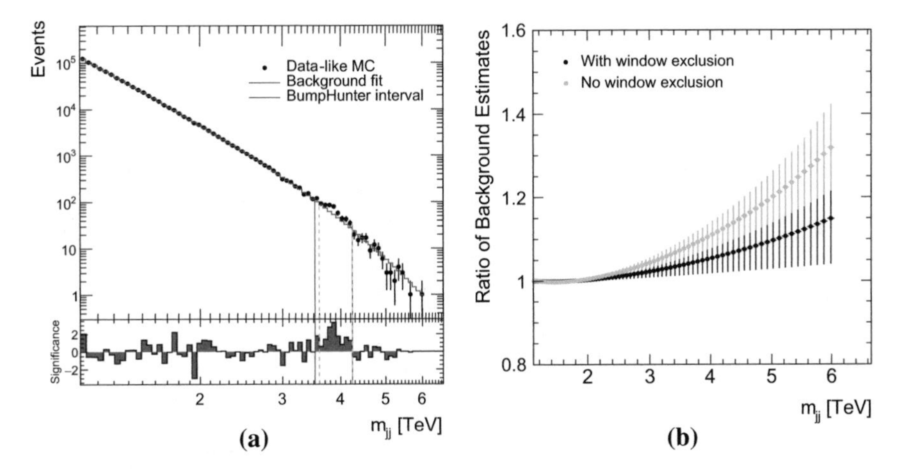

Fig. 6.9 These plots show the signal injection results for the high mass dijet analysis. **a** shows the combined data-like spectrum containing QCD and a 4 TeV q* signal in the black points. The background estimate is shown by the red curve, the green dashed lines indicate the window excluded in background estimate, and the blue vertical lines indicate the window with the largest significance. **b** shows the ratio between the data-like spectra with and without a signal present. The black points include a window exclusion in the determination of the background estimate, and the green points do not

6.4.2 *Identifying Signals*

For the high mass dijet analysis, the 3.6 fb^{-1} data-like MC spectrum was combined with 3.6 fb^{-1} data-like MC signal spectra, in order to simulate a data spectrum which contains a signal, allowing us to test the search procedure in the presence of a signal. Figure 6.9a shows the result of applying the search phase to a combined signal (4 TeV q*) plus background spectrum; the window excluded in order to determine the background estimation is indicated by the green dashed lines, and the most significant region identified by BumpHunter is indicated by the blue lines. The most significant region spans 4 TeV, the mass of the injected signal, and the global p-value for the selected bump is 0.0002, corresponding to 3.5σ significance. This indicates that the search procedure was able to successfully identify the signal at the correct mass.

Figure 6.9b shows the ratio between the background estimate obtained using the combined spectrum, and the background estimate obtained using the spectrum with no signal. The black points indicate the ratio with the window exclusion applied. Differences are expected between the two estimates, due to the window removal excluding bins used in the fitting procedure, resulting in a loss of information available to the fit. The ratio shows that at low masses the agreement between the two estimates is very good, and they only start to deviate as we reach the high mass tail region. For contrast, the ratio without the application of the window exclusion is shown in the green points. This indicates that the window exclusion is working well, as it is reducing the impact of the signal on the background estimate. Note that this is an

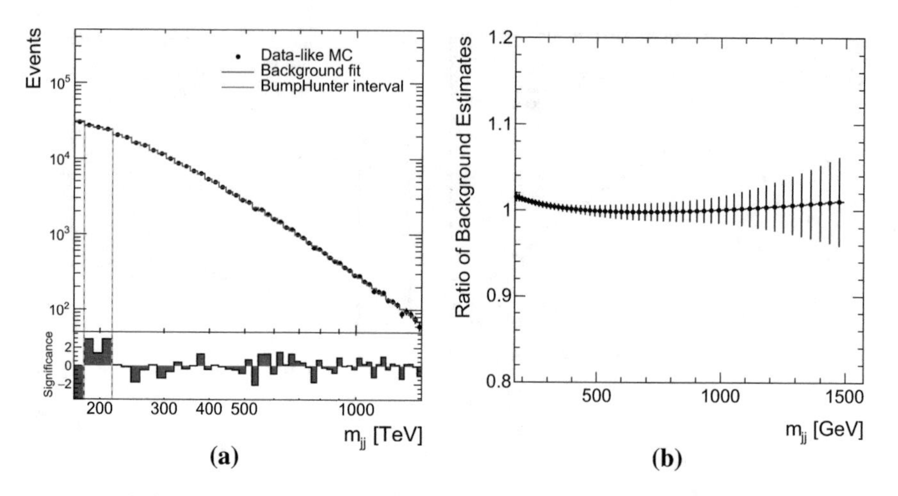

Fig. 6.10 These plots show the signal injection results for the dijet + γ analysis, using a combined data-like spectrum containing QCD and a 200 GeV Z' signal with $q_q = 0.3$

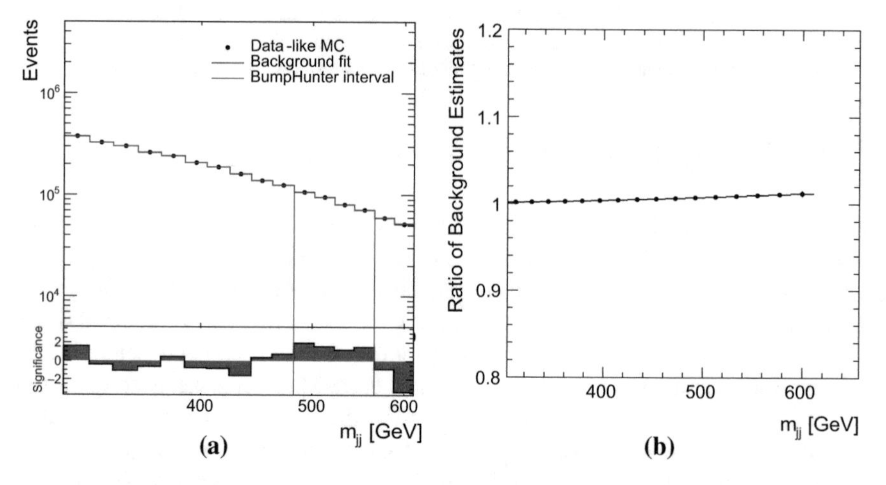

Fig. 6.11 These plots show the signal injection results for the dijet + jet analysis, using a combined data-like spectrum containing QCD and a 550 GeV Z' signal with $g_q = 0.3$

additional test which has been performed since the search procedure was applied to data.

The same tests were conducted for the dijet + ISR analyses. Note that there were insufficient MC events to produce data-like signals corresponding to the data luminosity, so scaled MC signal templates were used instead, with their statistical uncertainty set equal to the square root of the total number of events in the bin. Figure 6.10a, b show the results for the combined signal (200 GeV Z' with $g_q = 0.3$) plus background sample, for the dijet + γ analysis. Figure 6.11a, b show the results for the combined signal (550 GeV Z' with $g_q = 0.3$) plus background sample, for the dijet + jet analysis. In each case the most significant region selected by BUMPHUNTER corresponds to the injected signal mass. The global BUMPHUNTER p-value in each case is 0.0031 (2.7σ) and 0.0251 (1.96σ), for the dijet + γ analysis and the dijet + jet analysis, respectively. Figure 6.10a also illustrates that the search procedure is able to identify signals which occur close to the start of the fit range, and Fig. 6.11a illustrates that signals which occur close to the end of the fit range can be identified.

In Figures 6.10b and 6.11b the black points show the ratio between the background estimate obtained using the combined spectrum and the background estimate obtained using the spectrum with no signal. For the dijet + γ analysis, the ratio is close to one, indicating that the window exclusion is working well. The additional ratio without the application of the window exclusion is not shown because the fit could not converge without the application of the window exclusion, emphasising the importance of the window exclusion. For the dijet + jet analysis, a window exclusion was not applied as the global p-value was below the threshold for applying a window exclusion. The ratio shows that in the presence of this small excess the ratio remained close to one. Note that this is an additional test which has been performed since the search procedure was applied to data.

6.4.3 Spurious Signal Check

The background estimation procedure is tested for the introduction of spurious signals, i.e. excesses of events above the background estimate which are introduced by the fitting procedure and could be mistaken for signal. In order to investigate the size of spurious signals, MC spectra are scaled to the data luminosity and the uncertainties in each bin are set to $\frac{1}{\sqrt{N}}$. The background estimation procedure is then applied. Note that the same fit function range and number of parameters that were chosen for the background estimate in data were used for the spurious signal test for each analysis, with the exception of the dijet + jet analysis which utilised the four parameter fit function due to the results of the Wilks' test obtained when using MC.

The search procedure was performed on the MC scaled to the data luminosity. For each of the analyses, the BUMPHUNTER p-value was found to be above 0.9, indicating that no significant bumps were introduced by the background estimation procedure.

6.5 Search Results

The full search phase procedure, outlined in Sect. 6.3.2, was performed using the full dataset for each analysis. The goodness of fit for the fits to the data indicated by the χ^2 p-value was calculated to be 0.93 for the high mass dijet analysis, 0.58 for the dijet + γ analysis and 0.90 for the dijet + jet analysis, indicating that good quality fits were

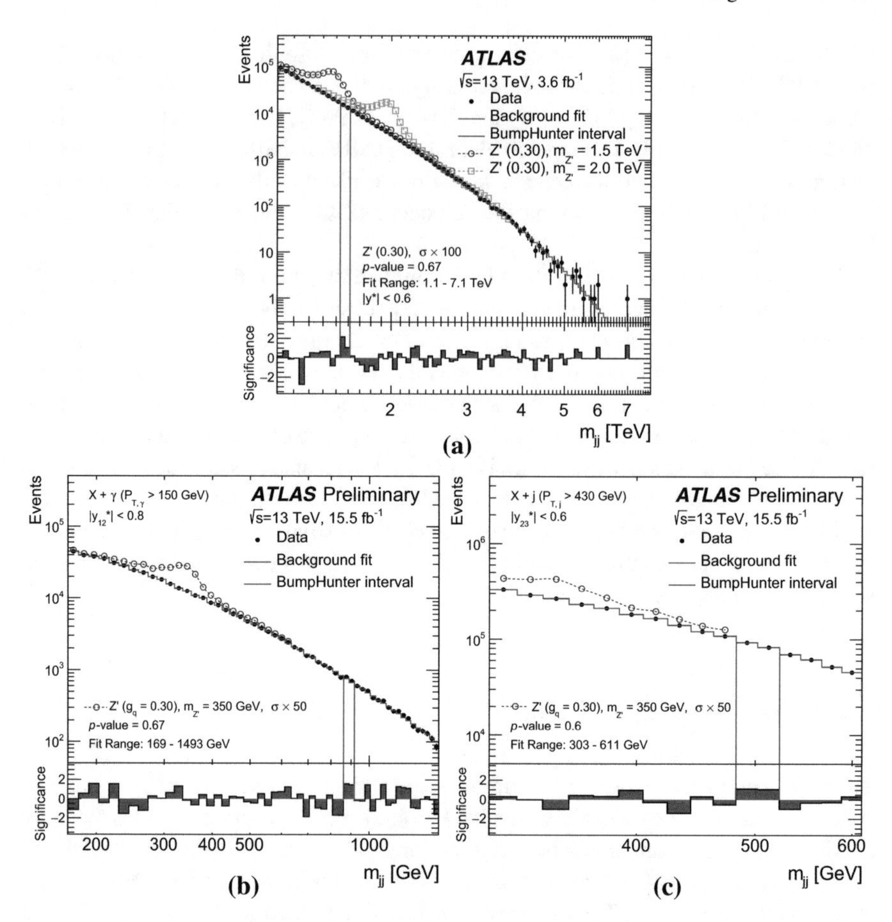

Fig. 6.12 The results of performing the search phase on the full dataset are shown for **a** the high mass dijet analysis, **b** the dijet + γ analysis, **c** the dijet + jet analysis. The top panel shows the data in black, the background estimate in red, the most discrepant region selected by BumpHunter by the vertical blue lines, and Z' MC signals ($g_q = 0.3$) with their cross-sections scaled by 100, 50 and 50, respectively, are overlaid with open circles. The lower panel shows the bin-by-bin significances

obtained. The results of the search are shown in Fig. 6.12, where the data points are indicated by the black points, the background estimate is shown by the red curve, and the blue vertical lines indicate the window with the largest significance found using BumpHunter. MC signals Z' ($q_q = 0.3$) with their cross-sections scaled by 100 in the high mass dijet analysis, and 50 in the dijet + ISR analyses, are overlaid with open circles. The lower panel shows the bin-by-bin significances for these distributions.

The window with the highest significance and the corresponding BumpHunter global p-value attained by each search are given in Table 6.1. These global p-values demonstrate that there is no evidence for localised excesses due to BSM phenomena being present in the mass spectra.

Table 6.1 Table showing the bump range and global p-value for each search

Analysis	Bump range [GeV]	Global p-value
High mass dijet	1533–1614	0.67
Dijet + γ	861–917	0.67
Dijet + jet	482–523	0.60

The plot of the local p-values for each mass window considered in the search is shown in Appendix D for each analysis. The distribution of the BUMPHUNTER test statistic from applying the same procedure to pseudo-experiments, together with the observed value from data, is also shown in Appendix D for each analysis.

References

1. Cranmer K (2011) Practical statistics for the LHC. In: Proceedings, 2011 European School of high-energy physics (ESHEP 2011). Cheile Gradistei, Romania, 7–20 Sept 2011. arXiv: 1503.07622 [physics.data-an]
2. Particle Data Group, Olive KA et al (2014) Review of particle physics. Chin Phys C38:090001. https://doi.org/10.1088/1674-1137/38/9/090001
3. Harris RM, Kousouris K (2011) Searches for dijet resonances at hadron colliders. Intl J Mod Phys A 26.30n31:5005–5055. https://doi.org/10.1142/S0217751X11054905, arXiv: 1110.5302 [hep-ex]
4. UA2 Collaboration (1991) A measurement of two-jet decays of the W and Z bosons at the CERN $\bar{p}p$ collider. Zeitschrift für Phys C Part Fields 49.1:17–28. https://doi.org/10.1007/BF01570793, ISSN: 1431-5858
5. CDF Collaboration (1995) Search for new particles decaying to dijets in $p\bar{p}$ Collisions at $\sqrt{s} = 1.8$ TeV. Phys Rev Lett 74:3538–3543. https://doi.org/10.1103/PhysRevLett.74.3538
6. ATLAS Collaboration (2015) Search for new phenomena in the dijet mass distribution using pp collision data at $\sqrt{s} = 8$ TeV with the ATLAS detector. Phys Rev D 91:052007. https://doi.org/10.1103/PhysRevD.91.052007, arXiv: 1407.1376 [hep-ex]
7. Fisher RA (1922) On the mathematical foundations of theoretical statistics. Philol Trans Roy Soc Lond A222:309–368. https://doi.org/10.1098/rsta.1922.0009
8. Aldrich J (1997) R. A. Fisher and the making of maximum likelihood 1912–1922. Stat Sci 12.3:162–176. http://www.jstor.org/stable/2246367, ISSN: 08834237
9. Pearson K (1900) FRS X. On the criterion that a given system of deviations from the probable in the case of a correlated system of variables is such that it can be reasonably supposed to have arisen from random sampling. Philos Mag Ser 5 50.302:157–175. https://doi.org/10.1080/14786440009463897
10. Cowan G (1998) Statistical data analysis. Oxford science publications, Clarendon Press, Oxford. ISBN 9780198501565
11. James F, Roos M (1975) Minuit - a system for function minimization and analysis of the parameter errors and correlations. Comput Phys Commun 10.6:343–367. https://doi.org/10.1016/0010-4655(75)90039-9, ISSN: 0010-4655
12. Wilks SS (1938) The large-sample distribution of the likelihood ratio for testing composite hypotheses. Ann Math Stat 9(1):60–62. https://doi.org/10.1214/aoms/1177732360
13. High mass dijet analysis team (2015) Private communication

14. Choudalakis G (2011) On hypothesis testing, trials factor, hypertests and the BumpHunter. In: Proceedings, PHYSTAT 2011 workshop on statistical issues related to discovery claims in search experiments and unfolding. Geneva. arXiv:1101.0390 [physics.data-an]
15. Choudalakis G (2011) Dijet searches for new physics in ATLAS. arXiv: 1109.2144 [hep-ex]
16. Albert A (2016) Searching for dark matter with cosmic gamma rays. IOP Concise Phys. Morgan & Claypool Publishers. http://iopscience.iop.org/book/978-1-6817-4269-4, ISBN: 9781681742694
17. Choudalakis G, Casadei D (2012) Plotting the differences between data and expectation. Eur Phys J Plus 127.2:25. https://doi.org/10.1140/epjp/i2012-12025-y, ISSN: 2190-5444
18. ATLAS Collaboration (2015) Dijet resonance searches with the ATLAS detector at 14 TeV LHC. Technical report, ATLAS-PHYS-PUB-2015-004. Geneva: CERN

Chapter 7
Limit Setting

The search results obtained in Chap. 6 indicated that there is no evidence for local excesses due to BSM phenomena in the mass spectra. The analyses proceed by using the data to derive limits on physical quantities for theoretical models of new physics. This enables us to quantify the phase space excluded by the analyses, allowing us to measure our progress and to compare our results with other experiments.

A theoretical introduction to limit setting is given in Sect. 7.1. A description of the systematic uncertainties considered in the limit setting for each of the analyses, and how they are incorporated, is given in Sect. 7.2. Section 7.3 describes how the limit setting is implemented for the analyses described in this thesis. In Sects. 7.4 and 7.5 the results of the limit setting are presented for benchmark models and for model-independent Gaussian shapes, respectively. Section 7.6 shows the impact of the analyses on the phase space for the Z' dark matter mediator model, and the exclusions achieved are compared to existing results from ATLAS and from other experiments.

7.1 Limit Setting Theoretical Background

As described in Chap. 6, the frequentist approach to probability treats the underlying parameters of a theory to be fixed, and the probability of obtaining the data under the assumption of the theory is calculated. In the Bayesian approach, the data are considered to be fixed, and a probability is assigned to the parameters of the theory [1]. A Bayesian approach was used to calculate the limits for the analyses in this thesis. Given the data, we assign probabilities to the number of signal events which could be present in our data, for several signal masses for each model of new physics. This allows us to obtain the maximum number of signal events which have not been excluded by the data. The condition for exclusion is based on a probability threshold, as explained in Sect. 7.1.2. By dividing the upper limit on the number of signal events

L. A. Beresford, *Searches for Dijet Resonances*, Springer Theses,
https://doi.org/10.1007/978-3-319-97520-7_7

by the luminosity of the dataset, we obtain the upper limit on the production cross-section of the signal σ, multiplied by the signal acceptance A, multiplied by the branching ratio to dijets BR, referred to as the *observed upper limit*. Note that the acceptance is defined as the fraction of events passing the analysis selection with respect to the total number of events. By comparing the observed upper limit to the theoretical production cross-section $\sigma \times \mathrm{A} \times \mathrm{BR}$ for the new physics model, ranges in signal mass can be excluded, corresponding to regions in which the observed limit is below the theoretical prediction. The probabilities utilised in the limit setting are calculated according to Bayes' Theorem [2]. A summary of the Theorem and its implications will be given here, for a detailed description see [3–5].

7.1.1 Bayes' Theorem

Bayes' theorem enables us to calculate the conditional probability of event A occurring, given that event B has occurred $P(A|B)$, and is given by

$$P(A|B) = \frac{P(B|A)P(A)}{P(B)}, \tag{7.1}$$

where $P(B|A)$ is the conditional probability of event B occurring, given that event A has occurred, $P(A)$ is the probability of event A occurring, and $P(B)$ is the probability of event B occurring. If we consider events where the outcomes are continuous rather than discrete, an equivalent form of Bayes' theorem exists with probabilities replaced by probability density functions, denoted by lower case p.

In the limit setting, we would like to obtain the probability density function for the parameters of our hypothesis, given the data. For each signal mass and new physics model, our hypothesis is the combination of the background estimate, and the Monte Carlo signal m_{jj} distribution (signal template). The background estimate is now obtained by performing a fit to the data distribution using the same function which was utilised in the search phase, combined with the signal template with floating signal normalisation, i.e. a signal plus background fit, to ensure that the background estimate does not incorporate the signal. An example of a hypothesis (background estimate plus signal template) is shown in Fig. 7.1, for a 5 TeV excited quark in the high mass dijet analysis.

The hypothesis is a function of several parameters. One of these parameters is the normalisation of the signal template ν, which is equivalent to the number of signal events; this is our parameter of interest. In addition to the parameter of interest, there are other parameters, referred to as nuisance parameters $\boldsymbol{\theta}$, which must be taken into account. In the analyses in this thesis, the nuisance parameters represent sources of systematic uncertainty on the hypothesis, and they can alter the shape and normalisation of the signal template and the background estimate, altering our hypothesis. Examples of nuisance parameters include the statistical uncertainty on

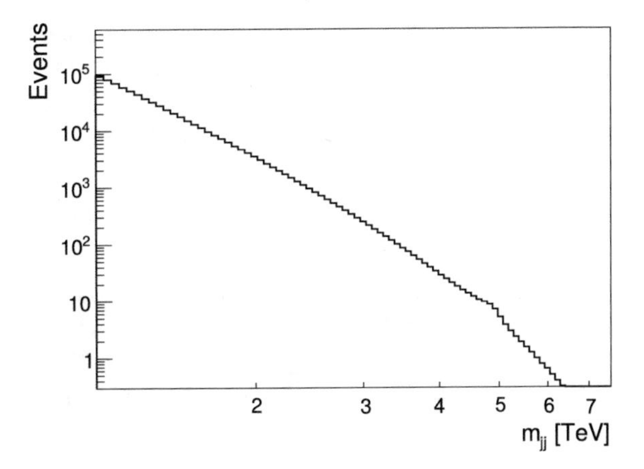

Fig. 7.1 This figure shows an example of a hypothesis in the high mass dijet analysis. The hypothesis consists of the background estimate, plus the signal template for a 5 TeV excited quark. This hypothesis is a function of several parameters, which can alter the shape and normalisation of the background estimate and the signal template, therefore altering the overall hypothesis

the background estimate, and the jet energy scale uncertainty. A description of all of the systematic uncertainties will be given in Sect. 7.2.

By utilising Bayes' theorem, given in Eq. (7.1), we can calculate the probability density function (p.d.f) for our hypothesis parameters (ν, $\boldsymbol{\theta}$), given the data. This p.d.f is referred to as the posterior, as it reflects our knowledge about the hypothesis parameters after analysing the data, and is given by the following equation:

$$p(\nu, \boldsymbol{\theta}|\text{Data}) = \frac{\mathcal{L}(\nu, \boldsymbol{\theta}|\text{Data})\pi(\nu, \boldsymbol{\theta})}{\text{p(Data)}}. \quad (7.2)$$

The likelihood of the hypothesis parameters, given the data, is denoted by $\mathcal{L}(\nu, \boldsymbol{\theta}|$ Data), and is equivalent to $p(\text{Data}|\nu, \boldsymbol{\theta})$. This means that we calculate the probability of obtaining our data for a given set of hypothesis parameters ν and $\boldsymbol{\theta}$ [6]. Further details about the likelihood will be given in Sect. 7.3. The prior probability density for the hypothesis parameters is denoted by $\pi(\nu, \boldsymbol{\theta})$. This is equivalent to $\pi(\nu) \prod_i \pi(\theta_i)$, as the parameter of interest ν, and each of the nuisance parameters θ_i, are independent from one another. The prior probability density for the signal normalisation $\pi(\nu)$, and the prior probability densities for the nuisance parameters $\pi(\theta_i)$ reflect our knowledge or belief about these parameters prior to analysing the data. For example, we might believe that there are zero signal events in our data, and hence, we might choose the signal normalisation prior $\pi(\nu)$ to peak at zero and to fall off with increasing ν. Alternatively, we might choose to express a lack of knowledge, and choose a flat prior which doesn't change with ν. The priors are selected by the analyser, and the choices for $\pi(\nu)$ and $\pi(\theta_i)$ will be described in Sects. 7.1.3, and 7.2, respectively. The probability density for obtaining the data is denoted by p(Data), and is equivalent to the integral of the numerator with respect to ν and θ_i. Hence, p(Data) ensures that the posterior $p(\nu, \boldsymbol{\theta}|\text{Data})$ is normalised to unity. The overall normalisation of the posterior does not affect the obtained upper limit, and therefore, this factor can be dropped in the limit setting.

In the limit setting, we are ultimately interested in obtaining the marginalised posterior $p(\nu|\text{Data})$ which is the p.d.f. for the number of signal events, given the data. In order to obtain $p(\nu|\text{Data})$, it is necessary to integrate the posterior $p(\nu, \boldsymbol{\theta}|\text{Data})$, given by Eq. (7.2), over the nuisance parameters $\boldsymbol{\theta}$, i.e.,

$$p(\nu|\text{Data}) = \int p(\nu, \boldsymbol{\theta}|\text{Data})d\boldsymbol{\theta}. \tag{7.3}$$

This multi-dimensional integral over the nuisance parameters is referred to as marginalisation. Details about the techniques used to perform the marginalisation will be given in Sect. 7.3. By substituting Eq. (7.2) into (7.3) utilising $\pi(\nu, \boldsymbol{\theta}) = \pi(\nu)\prod_i \pi(\theta_i)$, and dropping the normalisation factor $p(\text{Data})$, we obtain the following form of Bayes' equation:

$$p(\nu|\text{Data}) \propto \int \mathcal{L}(\nu, \boldsymbol{\theta}|\text{Data})\pi(\nu)\prod_i \pi(\theta_i)d\boldsymbol{\theta}. \tag{7.4}$$

This final form of Bayes' equation is utilised in the limit setting. Equation (7.4) can be interpreted as follows: the prior knowledge or belief of the analyser, encoded in the priors $\pi(\nu)$ and $\pi(\theta_i)$, is updated by the outcome of the experiment, encoded in the likelihood $\mathcal{L}(\nu, \boldsymbol{\theta}|\text{Data})$, in order to obtain the marginalised posterior $p(\nu|\text{Data})$. This is illustrated pictorially in Fig. 7.2, neglecting the presence of nuisance parameters for simplicity. The signal prior $\pi(\nu)$, shown in red, has been chosen to favour a non-zero number of signal events ν. However, the likelihood, shown in blue, indicates that the data favours zero signal events. The posterior, shown in black, is influenced by both the prior and the likelihood. Figure 7.2a shows the scenario for a smaller sample of data, and Fig. 7.2b shows the scenario for a larger sample of data. As data is added, the likelihood function becomes more dominant and the posterior becomes more similar to the likelihood function.

7.1.2 *Upper Limit*

By integrating the marginalised posterior distribution $p(\nu|\text{Data})$ across a given region, e.g. $\nu = 0$ to $\nu = 10$, we obtain the probability that the true value of ν lies in this region [3]. A common choice made in particle physics is to define the upper limit ν_{upper} as the value below which 95% of the marginalised posterior lies, referred to as the 95% credibility level (C.L.) [3]. This is expressed mathematically by the equation

$$0.95 = \int_{-\infty}^{\nu_{\text{upper}}} p(\nu|\text{Data})d\nu, \tag{7.5}$$

where $p(\nu|\text{Data})$ is given by Eq. (7.4). This means that we are 95% sure that the number of signal events that could be present in our data are equal to ν_{upper} or fewer.

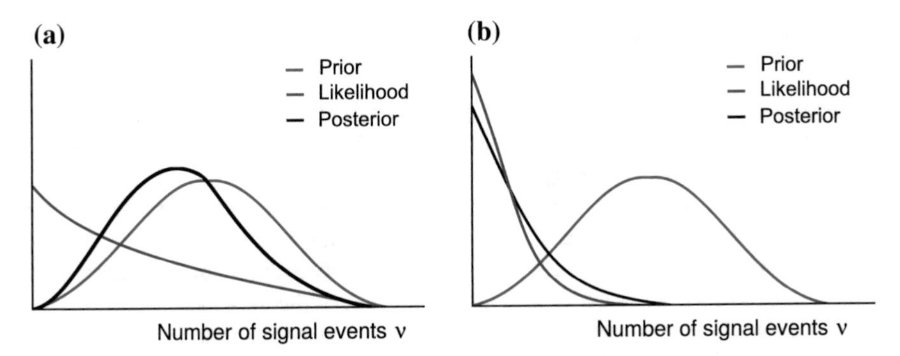

Fig. 7.2 This diagram illustrates the relationship between the prior, the likelihood and the posterior. The prior for the signal normalisation $\pi(\nu)$ is shown in red, the likelihood $\mathcal{L}(\nu|\text{Data})$ is shown in blue, and the posterior $p(\nu|\text{Data})$ is shown in black. Nuisance parameters have been neglected in this example for simplicity. Figure **a** shows the outcome for a smaller data sample, and Figure **b** shows the outcome for a larger data sample. Illustrating that for large data samples the likelihood can influence the posterior distribution more. This figure is based on [7]

7.1.3 *Choice of Signal Prior*

As previously mentioned, the prior for the hypothesis reflects the belief of the analyser, prior to analysing the data. The choice of priors for the nuisance parameters will be addressed in Sect. 7.2. In this section we discuss the prior for the parameter of interest $\pi(\nu)$. The prior is selected by the analyser, and it can be more or less informative. For example, the prior could be the posterior distribution from a previous experiment, or be based on the prior knowledge or belief of the analyser, making it *informative* [8]. Alternatively, an *uninformative* prior can be used to reflect ignorance about the value of the parameter, such that the prior has minimal influence on the obtained posterior.

An uninformative uniform prior was chosen for the analyses presented in this thesis. This is a simple and natural choice for the signal prior, and is consistent with previous dijet analyses, simplifying the comparison of results. Additionally, by minimising the influence that the prior has on the obtained posterior, we avoid introducing a large bias from the selection of an inappropriate informative prior [9]. Instead, we place our emphasis on the likelihood, which is determined using the data.

In order to be able to normalise $\pi(\nu)$ and to ensure that most of the distribution is not at infinity, a cut-off is defined for $\pi(\nu)$ [8]. This cut-off is set to a very large, but finite number of signal events ν_{max}, which is defined as the ν value corresponding to the position where the likelihood is a factor of 10^5 times smaller than its maximum value. The chosen prior is uniform and equal to $\frac{1}{\nu_{\text{max}}}$ in the range $[0, \nu_{\text{max}}]$ and is set to zero elsewhere, ensuring that, for example, negative numbers of signal events are not permitted.

7.2 Systematic Uncertainties

As previously mentioned, the nuisance parameters in the analyses described in this thesis correspond to sources of systematic uncertainty. The impact of the systematic uncertainties is accounted for by allowing the signal template and the background estimate to vary within their systematic uncertainties, altering the shape and normalisation of the hypothesis (the combination of the signal template and the background estimate). Alternatively, one could have chosen to apply the systematic uncertainties to the data, since we are making a comparison between the hypothesis and the data; however, for simplicity it was chosen to apply the systematics to the hypothesis.

The usual prior distribution selected to represent systematic uncertainties is the Gaussian prior. This prior gives decreased probability as we move further away from the nominal value, out to larger uncertainty values. All of the systematic uncertainties considered in the analyses described in this thesis utilise a Gaussian prior with a parameter range of $\theta \in \{-3\sigma, +3\sigma\}$, with the exception of the function choice uncertainty, which uses a Gaussian prior with a parameter range of $\theta \in \{0\sigma, +1\sigma\}$. Each of the systematic uncertainties will be defined in this section. The systematic uncertainties considered fall into two categories: uncertainties on the background estimate, which are described in Sect. 7.2.1, and uncertainties on the signal templates, which are described in Sect. 7.2.2.

7.2.1 *Uncertainties on the Background Estimate*

As previously mentioned, the background estimate in the limit setting is determined by performing a signal plus background fit to the data. For consistency, a signal plus background fit is utilised throughout the limit setting in the determination of the uncertainties associated with the background estimate. The uncertainties considered are as follows:

- **Statistical uncertainty on the fit**: An uncertainty on the best fit parameter values, due to the statistical uncertainty on the data.
- **Function choice**: An uncertainty on the choice of parametrisation.

The statistical uncertainty on the fit is given by the 1σ confidence interval in which the best fit parameter values would lie in 68% of repeated experiments [10]. In principle, this interval can be calculated using the covariance matrix for the fitted parameters; however, if the parameters of the fit are strongly correlated or there is a bound on a parameter of the fit, then the covariance matrix calculated by MINUIT can be unreliable [11]. To avoid these problems, the uncertainty was determined using pseudo-experiments. The best fit was found for 100 pseudo-experiments generated from the nominal fit to the data. The standard deviation of the function value for all the pseudo-experiments in each m_{jj} bin was calculated, and this defines the $\pm 1\sigma$

statistical uncertainty on the fit. In order to obtain the $+3\sigma$ statistical uncertainty, for example, the $+1\sigma$ uncertainty band is scaled up by a factor of 3.

The function choice uncertainty is determined by fitting the data with the nominal fit function and an alternative fit function with an additional degree of freedom. In each m_{jj} bin, the difference between these fits is scaled to the root mean square of the difference between nominal and alternate fits to 100 pseudo-experiments generated from the data itself. This defines the $+1\sigma$ function choice uncertainty. A Gaussian prior with a parameter range of $\theta \in \{0\sigma, +1\sigma\}$ was chosen for the function choice uncertainty, where 0σ corresponds to the nominal fit function, and $+1\sigma$ corresponds to the function choice uncertainty defined above. For values in between 0σ and $+1\sigma$, the difference between the function choice uncertainty and the nominal fit is scaled to the desired uncertainty, and the resulting distribution is then added to the nominal function.

Figure 7.3 shows the nominal fit to the data, the $\pm1\sigma$ statistical uncertainty on the fit, and the $+1\sigma$ function choice uncertainty for the high mass dijet analysis, the dijet + γ analysis, and the dijet + jet analysis. Note that for these figures a signal template was not utilised in the fit, in order to give an overview for all signal masses and models.

7.2.2 *Uncertainties on the Signal*

Several sources of systematic uncertainty are considered for the signal templates. The uncertainties have different effects on the signal template, for example, they can change their shape, or alter their normalisation. The sources of uncertainty and their impact on the signal templates is described below.

Jet Energy Scale and Resolution

As described in Sect. 4.2.6, there are more than 70 nuisance parameters associated with the jet energy scale calibration. The JetEtmiss performance group combines them to produce four strongly-reduced sets, with each set containing four nuisance parameters (three nuisance parameters for the high mass dijet analysis). In order to utilise a strongly reduced set, instead of the full set of nuisance parameters, each of the four sets must be tested, and the difference in the results obtained when using each set must be negligible. The impact was shown to be negligible for each of the analyses described in this thesis, and hence, one of the strongly reduced sets was used as the jet energy scale uncertainty.

The jet energy scale uncertainty affects the shape of the signal templates, as jets can be shifted in mass. It can also affect the acceptance for the signals. For each nuisance parameter in the set, signal templates are produced using jets shifted up and down by a given standard deviation σ. Templates are produced in steps of 0.5σ in the range $\{-3\sigma, +3\sigma\}$. For parameter values between two signal templates, the impact of the systematic is calculated by linearly scaling the difference in bin content between the two neighbouring templates to the desired σ value. For example, in order

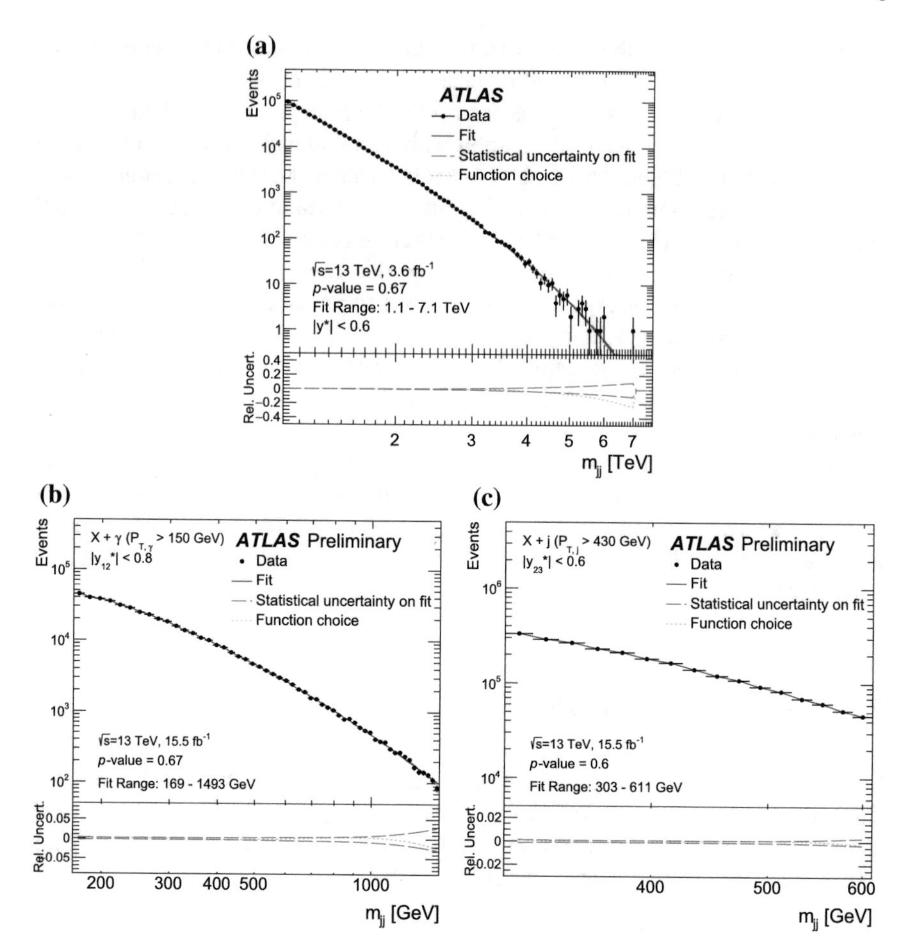

Fig. 7.3 The nominal fit to the data is shown in red, together with the statistical uncertainty on the fit, and the function choice uncertainty, shown by the blue dashed and dotted lines, respectively, for **a** the high mass dijet analysis, **b** the dijet + γ analysis, and **c** the dijet + jet analysis

to obtain the template corresponding to a shift of $+1.25\sigma$ for each bin, the difference between the $+1\sigma$ and $+1.5\sigma$ template is calculated and scaled by the shift in σ divided by the template separation in σ, i.e. 0.25/0.5; the bin-by-bin values are then added to the $+1\sigma$ template. By producing templates in steps of 0.5σ the range over which we must assume linear behaviour is reduced.

The same template method is also used for the jet energy scale uncertainty when setting limits on model-independent Gaussian shapes, rather than benchmark models. However, since the Gaussian signals are not produced using Monte Carlo, but are just simple Gaussian shapes, we must approximate the effect of the jet energy scale uncertainty on the templates. In order to do this, the relative difference between the peak position for a nominal benchmark signal sample and a benchmark signal sample with the jet energy scale uncertainty applied ($\pm 3\sigma$ for the high mass dijet analysis and $\pm 1\sigma$ for the dijet + ISR analyses). The value obtained is used to shift the Gaussian template in mass. For the high mass dijet analysis the q* signal was used, and for the dijet + ISR analyses the Z' signal with $g_q = 0.3$ was used. For the high mass dijet analysis the peak position was estimated using the mean, and the dominant jet energy scale nuisance parameter was used to calculate the uncertainty. This resulted in mass dependent shifts of up to 9% at high mass. For the dijet + ISR analyses, the peak position was estimated by fitting the signal templates with a Crystal ball function [12], and the quadrature sum of each of the nuisance parameters was used for the uncertainty. This resulted in a flat shift of ~2% for the dijet + ISR analyses.

The jet energy resolution uncertainty was described in Sect. 4.2. Now we must assess the impact of this uncertainty. In order to do this, the energy of each of the jets in our signal template is scaled by a *smearing factor*. The smearing factor for each individual jet is calculated by drawing a number from a Gaussian with width equal to the 1σ jet energy resolution uncertainty [13]. The relative difference in acceptance between the template with the JER uncertainty applied and the nominal template is then calculated in a ±50 GeV signal window around the nominal signal mass. This resulted in a flat 2% uncertainty on acceptance for the dijet + γ analysis and a flat 1% uncertainty on acceptance for the dijet + jet analysis, applied as a change to the normalisation of the signal template. For the high mass dijet analysis, an uncertainty on the jet energy resolution uncertainty was not included as it was deemed to be negligible, based on the studies performed by the ATLAS Collaboration for the dijet resonance search performed using $\sqrt{s} = 8$ TeV data [5, 14].

Luminosity

The determination of the integrated luminosity involves performing beam-separation scans. There are many sources of systematic uncertainty associated with the luminosity determination, for example, the alignment of the beam. A full description of the sources of systematic uncertainty and the determination of the overall systematic uncertainty is given in [15, 16]. The $\pm 1\sigma$ uncertainty on the integrated luminosity was determined to be ±9% for the high mass dijet analysis, using beam-separation scans performed in June 2015. The $\pm 1\sigma$ uncertainty for the dijet + ISR analyses was determined to be ±2.9%, using beam-separation scans performed in August 2015 and May 2016 which reduced the uncertainty. The luminosity uncertainty changes the normalisation of the signal template. For example, the $+1\sigma$ signal template corresponds to the nominal signal template scaled up by 9% for the high mass dijet analysis.

Parton Distribution Function

In order to generate the signal templates, a parton distribution function (PDF) must be utilised. There are uncertainties associated with the derivation of the PDF. For the analyses in this thesis, the PDF set from the NNPDF group [17] was utilised. In addition to providing the PDF, an ensemble of PDFs is also provided to allow the user to derive the PDF uncertainty [18]. The LHAPDF software [19] was used to reweight signal samples to each member of the ensemble. The uncertainty is given by the standard deviation of the signal acceptance when using each member in the ensemble [18]. A flat 1% uncertainty was assigned for the dijet + γ and dijet + jet analyses. For the high mass dijet analysis a flat 1% uncertainty was also used. This uncertainty was assigned based on the studies performed by the ATLAS Collaboration for the dijet resonance search performed using $\sqrt{s} = 8$ TeV data [14], in which the uncertainty was calculated by comparing the acceptance obtained using two different PDFs; one from the MSTW group [20] and one from the NNPDF group, as described in [5]. The PDF uncertainty changes the normalisation of the signal template. For example, the $+1\sigma$ signal template corresponds to the nominal signal template scaled up by 1% in each of the analyses. Note that the PDF uncertainty is not applied for the limits set on Gaussian shapes, as these are not generated using Monte Carlo.

Photon Identification, Energy Scale and Resolution

For the dijet + γ analysis, an additional uncertainty was required to account for changes in acceptance due to the photon identification, energy scale and resolution. The uncertainties are provided by the EGamma performance group and are described in [21, 22]. In each case, the relative change in signal acceptance in a ± 50 GeV signal window around the nominal signal mass was calculated after varying the signal template by $+1\sigma$. The combination of the three uncertainties resulted in a flat 3% uncertainty on the acceptance. The uncertainty changes the normalisation of the signal template, for example, the $+1\sigma$ signal template corresponds to the nominal signal template scaled up by 3% for the dijet + γ analysis.

Table 7.1 gives a summary of the systematic uncertainties considered in each analysis, and their size in percent. Note that the values for the statistical uncertainty on the fit, the function choice uncertainty, and the jet energy scale uncertainty vary with jet p_T or m_{jj}, and only a single value is displayed in the table for reference. The variation of the fitting uncertainties with m_{jj} is shown in Fig. 7.3, and the dependence of the jet energy scale uncertainty with jet p_T is shown in Fig. 4.15.

7.3 Limit Setting Implementation

The Bayesian Analysis Toolkit (BAT) [4] is used to obtain the marginalised posterior distribution $p(\nu|\text{Data})$. The analyser must provide BAT with a list of parameters ν and $\boldsymbol{\theta}$ and their corresponding prior distributions, as well as the likelihood function $\mathcal{L}(\nu, \boldsymbol{\theta}|\text{Data})$. The likelihood function is given by the product of the Poisson probability in each bin:

$$\mathcal{L}(\nu, \boldsymbol{\theta}|\text{Data}) = \prod_{\text{i}=1}^{\text{N}} \frac{\text{E}_\text{i}^{\text{D}_\text{i}} \text{e}^{-\text{E}_\text{i}}}{\text{D}_\text{i}!}, \tag{7.6}$$

Table 7.1 This table summarises the source and size of the systematic uncertainties applied in each of the analyses described in this thesis. It also indicates whether the systematic is applied to the background estimate or to the signal template. Note that the statistical uncertainty on the fit, the function choice uncertainty and the jet energy scale uncertainty displayed are for a single value of m_{jj} or jet p_T as indicated in parentheses

Systematic	Applied	High mass	Dijet + γ	Dijet + jet
Statistical	Bkg	10% (m_{jj} 7 TeV)	3% (m_{jj} 1.5 TeV)	0.5% (m_{jj} 600 GeV)
Function choice	Bkg	25% (m_{jj} 7 TeV)	2.5% (m_{jj} 1.5 TeV)	0.5% (m_{jj} 600 GeV)
Jet energy scale	Signal	3% (p_T 2.5 TeV)	2.3% (p_T 2.5 TeV)	2.3% (p_T 2.5 TeV)
Luminosity	Signal	9%	2.9%	2.9%
PDF	Signal	1%	1%	1%
JER	Signal	–	2%	1%
γ (ID, scale, reso.)	Signal	–	3%	–

where the product runs over all the bins i in the spectrum, E_i is the total number of expected events (signal + background, which depend on ν, $\boldsymbol{\theta}$) in bin i, and D_i is the number of data events in bin i. Note that the nominal signal template is initially normalised to unity, such that ν corresponds to the number of signal events. Template based systematics on the signal are scaled by the integral of the original nominal signal template, hence, they take into account changes in acceptance. For a given set of parameter values ν and $\boldsymbol{\theta}$, the expected number of events E_i in bin i is calculated by first applying the template based systematic uncertainties (the statistical uncertainty on the fit, the function choice and the jet energy scale uncertainty) to the signal or background distribution, accordingly, as these can modify the shape of the signal and background distributions. The signal distribution is then scaled by the parameter of interest (the signal normalisation) and the systematic uncertainties which can affect the signal normalisation, i.e. the JER, Luminosity, PDF, and the photon uncertainties for the dijet + γ analysis. The signal and background distributions are then added together, forming the expected spectrum, from which the bin-by-bin total number of expected events E_i are extracted. We are now able to calculate the likelihood for a given set of parameter values (ν, $\boldsymbol{\theta}$).

Now that the likelihood function, parameters and priors have been defined, the next step is to map out the posterior distribution $p(\nu, \boldsymbol{\theta}|\text{Data})$ (from which the marginalised posterior distribution $p(\nu|\text{Data})$ can be derived). Recall that the posterior is given by $p(\nu, \boldsymbol{\theta}|\text{Data}) = \mathcal{L}(\nu, \boldsymbol{\theta}|\text{Data})\pi(\nu)\prod_i \pi(\theta_i)d\boldsymbol{\theta}$, and hence, is a function of the parameters ν the signal normalisation and $\boldsymbol{\theta}$ the nuisance parameters representing sources of systematic uncertainty. In order to map out the posterior, we need to sample the parameter space of all allowed values of ν and $\boldsymbol{\theta}$ (ν is limited to the range 0 to ν_{max} defined in Sect. 7.1.3, and each θ_i parameter is limited to $\pm 3\sigma$, except for the parameter corresponding to the fit function choice uncertainty which is limited to 0–3σ). In order to sample this space efficiently BAT employs Markov Chain Monte Carlo (MCMC) [23, 24]. For full details about the MCMC implementation in BAT see [25, 26], an overview will be provided here. The basic idea is that we perform a random walk in parameter space (ν, $\boldsymbol{\theta}$), spending more time in regions of high probability density, i.e. sampling proportional to the posterior. Each set of parameters only depends on the previous set, not on the full history, hence, the sequence of parameter values is a Markov Chain [27]. The Metropolis-Hastings algorithm [23, 28] is used to generate the Markov Chains used in BAT, proceeding as follows:

1. The chain starts at position $\boldsymbol{x_1}$ in parameter space.
2. A new position $\boldsymbol{x_2}$ is proposed by individually selecting each new parameter from a Breit–Wigner distribution centered on the corresponding parameter in $\boldsymbol{x_1}$.
3. A random number r between 0 and 1 is selected from a uniform distribution
4. The value of the posterior $p(\nu, \boldsymbol{\theta}|\text{Data}) = \mathcal{L}(\nu, \boldsymbol{\theta}|\text{Data})\pi(\nu)\prod_i \pi(\theta_i)d\boldsymbol{\theta}$ for each set of parameters $\boldsymbol{x_1}$ and $\boldsymbol{x_2}$ is calculated, i.e. $p(\nu, \boldsymbol{\theta}|\text{Data})_1$ and $p(\nu, \boldsymbol{\theta}|\text{Data})_2$.
5. If $r < \frac{p(\nu,\boldsymbol{\theta}|\text{Data})_2}{p(\nu,\boldsymbol{\theta}|\text{Data})_1}$, we transition to the new position $\boldsymbol{x_2}$ and it is added to the chain, otherwise we remain at position $\boldsymbol{x_1}$ and it is added to the chain.
6. The process is then repeated with the chosen position defined as position $\boldsymbol{x_1}$.

An illustration of this procedure for two parameters θ_1 and θ_2 is shown in Fig. 7.4.

We have now performed a random walk in parameter space, with the chain preferentially transitioning to positions corresponding to high probability regions of the posterior [29, 30]. In this way we have mapped out the posterior distribution. By plotting the frequency of occurrence for individual parameters and normalising the distribution to unity, we then have access to the marginal posteriors for the parameter of interest $p(\nu|\text{Data})$, and for all of the nuisance parameters. A simplified example is given for the parameter of interest ν. Consider one chain and one parameter, for example, $\nu = 0, 1, 1, 0, 2, 0$. We would get a histogram of the marginalised posterior with three entries for $\nu = 0$, two entries for $\nu = 1$, and one entry for $\nu = 2$. This distribution is then normalised to unity to obtain the final marginalised posterior.

7.4 Model Dependent Limits

For each signal model and mass point, the posterior $p(\nu|\text{Data})$ is calculated as described in Sect. 7.3. An example of the posterior distribution for a 5 TeV q* signal in the high mass dijet analysis is shown in Fig. 7.5. The 95% quartile is indicated by the line on the plot. As previously described in Sect. 7.1.2, this defines the upper limit on the number of signal events. In addition to obtaining the posterior for the parameter of interest ν, we also obtain the posterior distributions for each of the nuisance parameters. As a cross-check of the limit setting procedure, we compare the posterior distributions obtained for each nuisance parameter to the prior used for that nuisance parameter. Examples of the comparison of the priors and posteriors are shown in Appendix E.

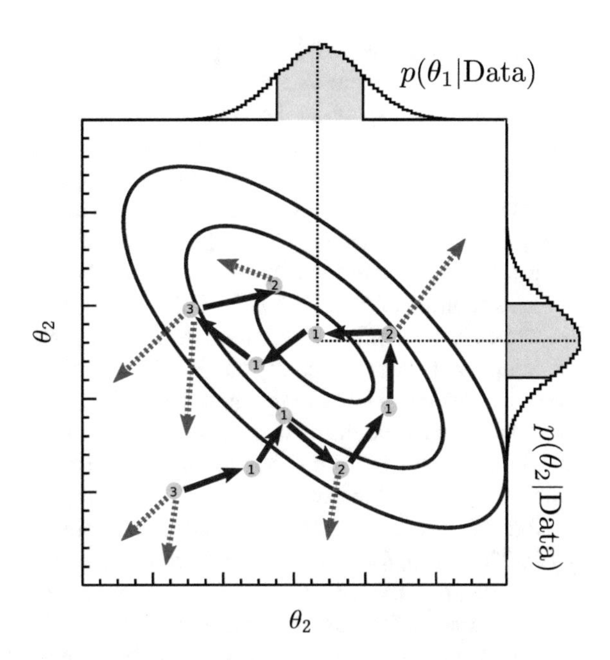

Fig. 7.4 This figure illustrates a random walk in parameter space (θ_1,θ_2). The numbers indicate the number of iterations the chain remained at this point in parameter space, the solid arrows indicate accepted transitions, and the dashed arrows indicate rejected transitions. The marginalised posterior distributions obtained for the two parameters $p(\theta_1|\text{Data})$ and $p(\theta_2|\text{Data})$ are also shown, and the shaded bands correspond to the central 68% of the distributions. Figure adapted from [25]

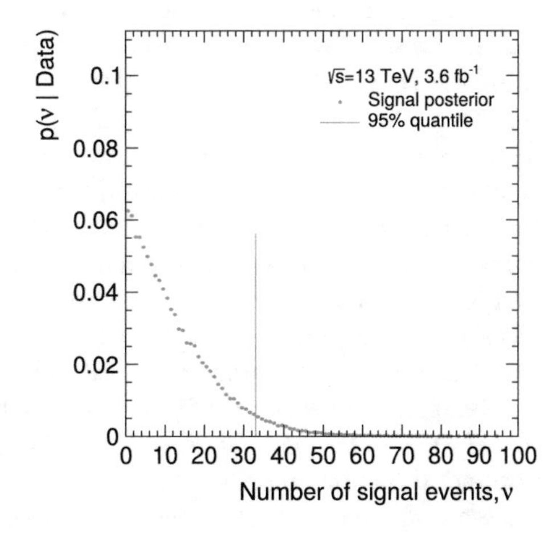

Fig. 7.5 The marginalised posterior $p(\nu|\text{Data})$ as a function of the number of signal events ν, for the 5 TeV q* mass point in the high mass dijet analysis. The line indicates the 95% C.L. upper limit on ν

In order to produce the final limit plot, for each signal mass point the upper limit on the number of signal events is divided by the luminosity of the data set. Since only events which fall within our analysis selection are considered in the limit setting, and only dijet final states are utilised, the final limit is on the production cross-section of the signal σ, multiplied by the signal acceptance A, multiplied by the branching ratio to dijets BR. The upper limit on $\sigma \times \text{A} \times \text{BR}$ is then displayed as a function of signal mass. In addition to displaying the observed limit, the theoretical prediction for $\sigma \times$

A $\times$ BR is also displayed. By comparing the theoretical prediction to the observed limit curve we can exclude ranges in mass for the model we are setting limits on.

In addition to displaying the observed upper limit on $\sigma \times$ A $\times$ BR, the *expected upper limit* curve and 1σ and 2σ uncertainty bands are also displayed. The expected upper limit is produced by setting the nuisance parameters to their maximum-likelihood values, with ν set equal to zero. This spectrum is then used to produce pseudo-experiments, generated in the same manner as described at the start of Chap. 6. The limit setting procedure is then performed for each pseudo-experiment, and a distribution of their 95% C.L. limits on $\sigma \times$ A $\times$ BR is produced. The median, 1σ, and 2σ values of this distribution define the expected limit curve and the corresponding 1σ and 2σ uncertainty bands. The final limit plots obtained for each of the signal models considered in the high mass dijet analysis are shown in Fig. 7.6. It is worth noting that the limits weaken in areas of the mass spectrum which displayed an excess of events, this can be seen more clearly in the model-independent limits which are presented in Sect. 7.5.

Note that, for the quantum black hole models, the limit is set on $\sigma \times$ A, not $\sigma \times$ A $\times$ BR. The reason for this is that all decays of the quantum black holes are simulated when generating the MC, not just the decays to two partons, so by calculating the acceptance as the fraction of events passing the analysis selection with respect to the total number of events, the branching ratio is taken into account in the acceptance.

The upper left quadrant of Fig. 7.6 shows the limits set on the quantum black hole models for the three different scenarios: the ADD quantum black hole generated using the BLACKMAX generator (BM), the ADD quantum black hole generated using the QBH generator (QBH) and the RS quantum black hole generated using the QBH generator (RS). A single observed and expected limit are displayed, as the signal shape is very similar for the three scenarios. Figure 7.7a shows a comparison between the signal shape for a 6 TeV quantum black hole for each of the three scenarios. Three separate theoretical predictions are displayed as each scenario has a different predicted cross-section. Additionally, Fig. 7.7a shows that the quantum black hole models result in signal shapes without large tails at high or low mass, enabling us to set very strong observed and expected limits.

The upper right quadrant of Fig. 7.6 shows the limits set on the Z' model with $g_q = 0.3$. The lower left quadrant shows the limits set on the q* model, and the lower right quadrant shows the limits set on the W' model. The observed limits for low signal masses is similar for each of the three signal models. This is because their signal shapes are similar at low mass, as shown in Fig. 7.7b. However, at higher masses, the limits on the W' signal weaken with respect to the limits on q* signal, due to the large low mass tails of the W' signal.

The strength of the mass limits set depends on both the strength of the observed limit, and the theoretical prediction. A summary of the mass limits achieved in the high mass dijet analysis are given in Table 7.2, together with the results obtained by the dijet resonance search performed by the ATLAS Collaboration using $\sqrt{s} =$ 8 TeV data [14], for reference. The 13 TeV observed limit increased the exclusion in mass by up to 2.6 TeV for the quantum black hole models, by more than 1 TeV for

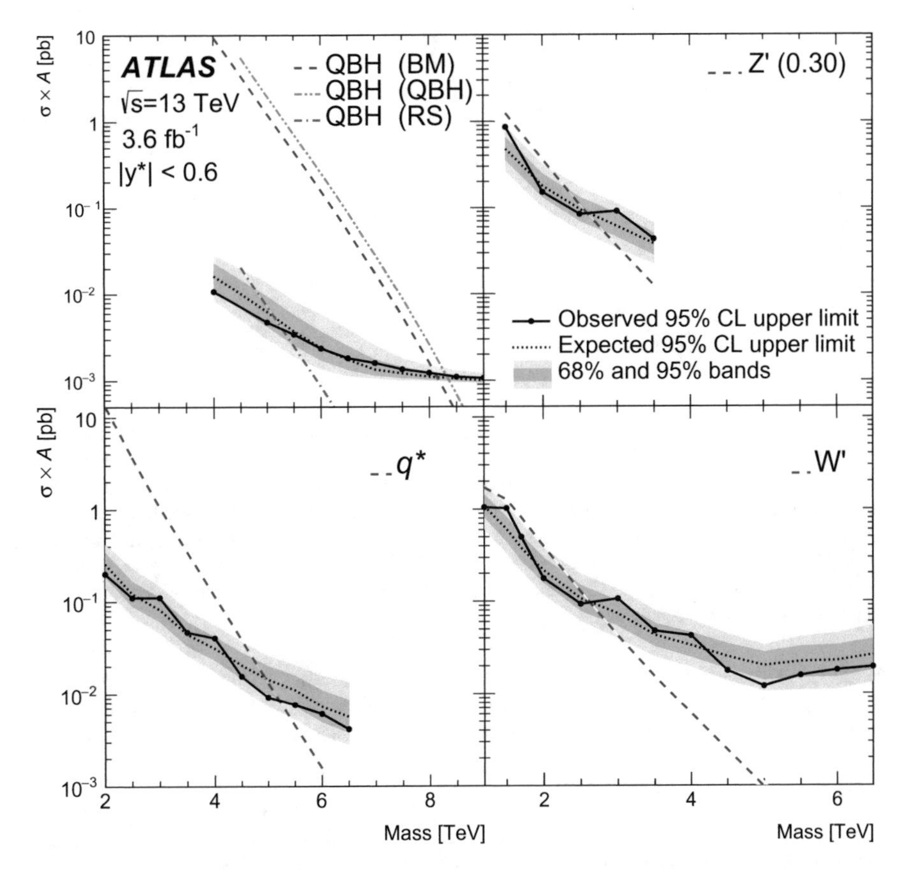

Fig. 7.6 The upper limit set on $\sigma \times \text{A} \times \text{BR}$ at 95% C.L. as a function of signal mass for the four signal models considered in the high mass dijet analysis. Clockwise from the upper left hand quadrant, the limits for the quantum black hole models, the Z' model with a coupling to quarks of $g_q = 0.3$, the W' model, and the excited quark model are shown. The observed limit is shown by the solid black curve, the expected limit is shown by the dotted black curve, and the corresponding 1σ and 2σ uncertainty bands are shown in green and yellow, respectively. The theoretical prediction is shown by the dashed blue line. Note that the axis is labelled as $\sigma \times \text{A}$ due to the inclusion of the limits on quantum black hole models, for which the branching ratio is included in the acceptance, as described in the text. For all other models the limit is on $\sigma \times \text{A} \times \text{BR}$

the excited quark model and by 0.1 TeV for the W' model, with respect to the 8 TeV observed limit.

The limits achieved for the Z' model in the high mass dijet analysis are presented in Fig. 7.8. The limits are shown in the plane of the Z' coupling to quarks g_q, versus the Z' mass $M_{Z'}$. For each mass and coupling point considered, the ratio between the observed limit and the theoretical prediction is shown. Mass and coupling points which are shown in blue or white have a ratio which is less than one, indicating that the point is excluded at 95% C.L.. The figure indicates that Z' signals with masses less than or equal to 2.5 TeV with couplings equal to or exceeding $g_q = 0.3$ are

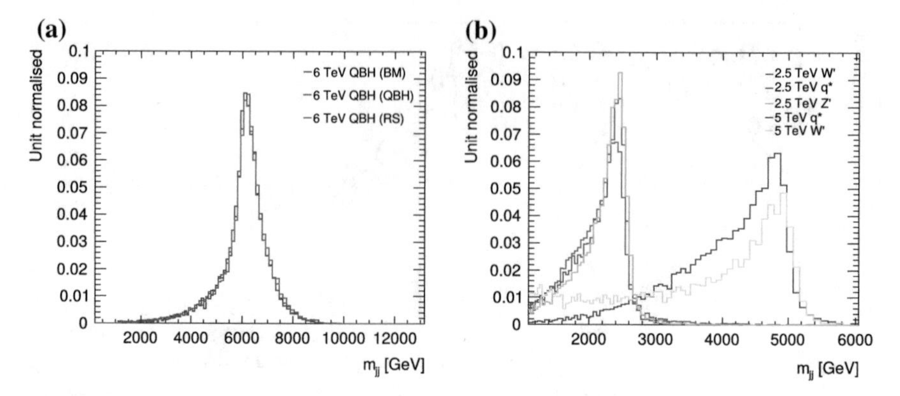

Fig. 7.7 A comparison of the shapes of the signal templates utilised in the high mass dijet analysis. Note that the areas of the signal templates have been normalised to unity. In Figure **a** 6 TeV quantum black hole signal templates are compared for three scenarios: an ADD quantum black hole produced using the BLACKMAX generator (BM), an ADD quantum black hole produced using the QBH generator (QBH), and an RS quantum black hole produced using the QBH generator (RS). In Figure **b** the comparison is made between a 2.5 TeV W' signal, q* signal and Z' signal, indicating their similarity in signal shape at low mass. In contrast, the 5 TeV q* signal and W' signal are seen to have very different signal shapes

Table 7.2 This table shows the signal model, the 95% C.L. exclusion limit on mass achieved by the 8 TeV high mass dijet resonance search (for reference), and the 13 TeV high mass dijet resonance search observed and expected limits. Masses below those shown for each signal model are excluded at 95% C.L.

Model	95% C.L. Exclusion limit		
	Observed 8 TeV	Observed 13 TeV	Expected 13 TeV
Quantum black holes, ADD (BLACKMAX generator)	5.6 TeV	8.1 TeV	8.1 TeV
Quantum black holes, ADD (QBH generator)	5.7 TeV	8.3 TeV	8.3 TeV
Quantum black holes, RS (QBH generator)	–	5.3 TeV	5.1 TeV
Excited quark	4.1 TeV	5.2 TeV	4.9 TeV
W'	2.5 TeV	2.6 TeV	2.6 TeV

excluded. The results presented in this figure are used in the production of the dark matter summary plot, which will be explained in detail in Sect. 7.6.

For the dijet + ISR analyses, limits are set on the Z' model for various mass and coupling points. Examples of the limits set on $\sigma \times A \times BR$ are shown in Fig. 7.9 for the Z' model with $g_q = 0.3$, for both the dijet + γ analysis and the dijet + jet analysis.

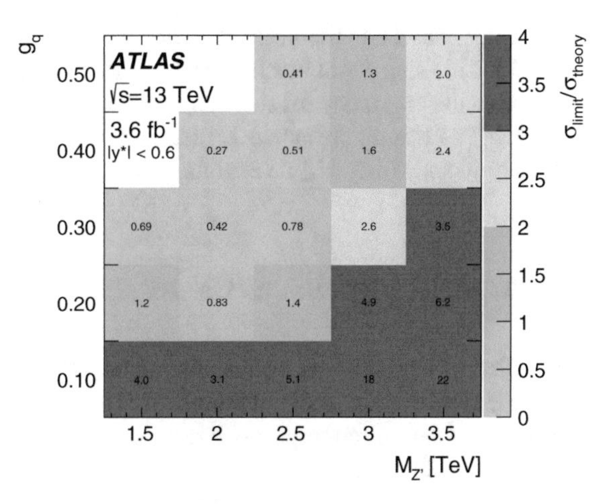

Fig. 7.8 In the plane of the coupling of the Z' to quarks g_q, versus the Z' mass $M_{Z'}$, the ratio between the observed limit and the theoretical prediction is indicated by the number. The mass and coupling points with a ratio less than one are excluded at 95% C.L., these points are indicated by a blue or white box

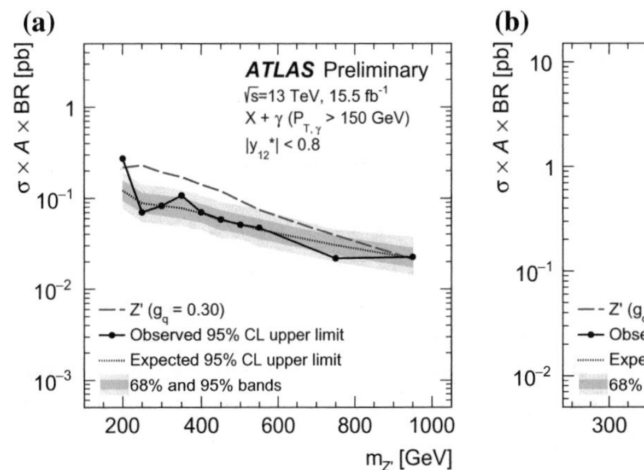

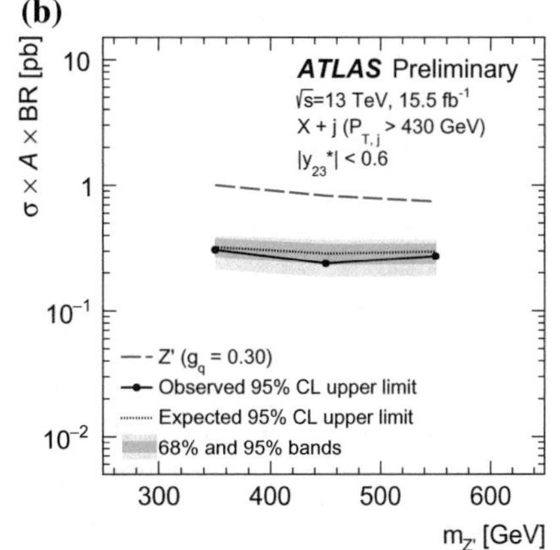

Fig. 7.9 The upper limit set on $\sigma \times \text{A} \times \text{BR}$ at 95% C.L. as a function of the Z' mass $m_{Z'}$ is shown for the Z' model with a coupling to quarks of $g_q = 0.3$ for **a** the dijet + γ analysis, and **b** the dijet + jet analysis. For the dijet + γ analysis, each of the curves (observed, expected and the theoretical prediction) have been corrected for experimental inefficiencies through division by the efficiency values shown in Fig. 5.13, for each signal mass point and coupling

By comparing the observed limit to the theoretical prediction, we can see that for the dijet + γ analysis, the observed curve lies below the theoretical prediction for nearly all mass points, with the exception of the 200 GeV mass point and the 950 GeV mass point. The mass points for which the observed curve lies below the theoretical curve are considered to be excluded at 95% C.L.. For the dijet + jet analysis, we see that all the mass points considered are well below the theoretical prediction.

For the dijet + γ analysis, an additional step was required in order to obtain the final limits. In order to display the limit on $\sigma \times \text{A} \times \text{BR}$, the efficiency for each mass and coupling point must be divided out for the observed limit, expected limit, uncertainty bands, and theoretical prediction. The efficiency was given previously in Fig. 5.13. This step is not needed for the high mass dijet analysis and the dijet + jet analysis as the efficiency for reconstructing jets is 100% in the phase space of these analyses. The results presented here are used in the production of the dark matter summary plots, which will be explained in detail in Sect. 7.6.

7.5 Model-Independent Limits

In addition to setting limits on specific models of new physics, limits are also set for generic Gaussian shaped signal templates. These limits are very useful for theorists, as it allows them to reinterpret our results to set limits on their own signal models, enabling them to determine which regions of phase space are excluded by our results.

Gaussian signal templates are produced with a range of different widths, with the narrowest Gaussian signal template having a width-to-mass ratio matching the fractional dijet mass resolution of the detector. Limits are then set on the Gaussian shapes in the same way as for the model dependent limits. For the dijet + γ analysis, the efficiency is corrected for in the same manner as for the model dependent limits; however, a flat efficiency value of 0.81 is utilised; this is the average efficiency for each mass and coupling value.

The limits set on the Gaussian signals are shown in Fig. 7.10 for each of the analyses described in this thesis. It is seen that in general, the narrowest Gaussian signals considered, shown in red, tend to set the strongest limits. They also tend to be the most sensitive to the statistical fluctuations in the data, resulting in less smooth limit curves. As previously mentioned, the limits weaken in areas of the mass spectrum which displayed an excess of events. This can be observed, for example, in the limits for the dijet + γ analysis, where the most significant mass region selected by BUMPHUNTER is 861–917 GeV, and it is observed that the Gaussian limits weaken in this mass range. Note that limits are only shown for Gaussian signal templates which are at least two times the width of the Gaussian σ_G from the edge of the fit range. Instructions for the reinterpretation of the Gaussian limits can be found in Appendix A of [14], and tables giving the precise value of the limit for each mass point and width are given in Appendix F.

7.6 Summary of Limits

The results from setting limits on the Z' model in all three analyses are used in the production of dark matter summary plots. The summary plot, showing 95% C.L. limits on the coupling of the Z' to quarks g_q, versus the Z' mass $m_{Z'}$ is shown in Fig. 7.11. The observed limits are shown by the solid lines, and the expected limits are shown by the dotted lines. Coupling values above the observed limits are excluded. The limits from the high mass dijet analysis are shown in dark blue, the limits from

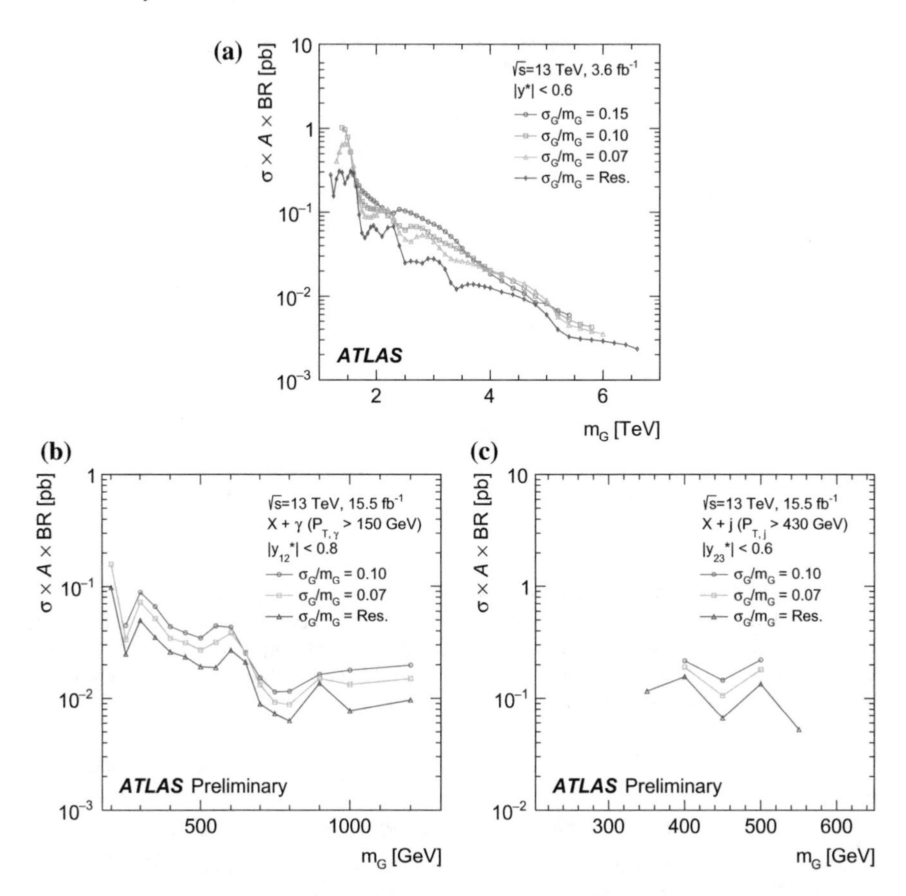

Fig. 7.10 The upper limit set on $\sigma \times A \times BR$ at 95% C.L. as a function of the mass of the Gaussian signal m_G is shown for **a** the high mass dijet analysis, **b** the dijet + γ analysis, and **c** the dijet + jet analysis. The observed limits shown in red correspond to very narrow Gaussians with a width-to-mass ratio equal to the fractional dijet mass resolution (Res.). Limits set on Gaussians with higher width-to-mass ratios are shown in other colours. For the dijet + γ analysis each of the limits has been corrected for experimental inefficiencies through division by 0.81, the average photon efficiency

the dijet + γ analysis are shown in red and the limits from the dijet + jet analysis are shown in purple. Additionally, the limits from the Trigger-object Level Analysis (TLA) [31], not described in this thesis, are shown in blue.

Figure 7.11 illustrates how the dijet searches are working together in order to exclude phase space for the Z' model. The dijet + γ analysis allows limits to be set down to 200 GeV in mass, and sets limits over a larger range in mass; however, in the region that the dijet + jet analysis covers, it sets more stringent limits on g_q. The reason for this is due to the higher theoretical prediction for the dijet + jet case, as shown in Figure 7.9.

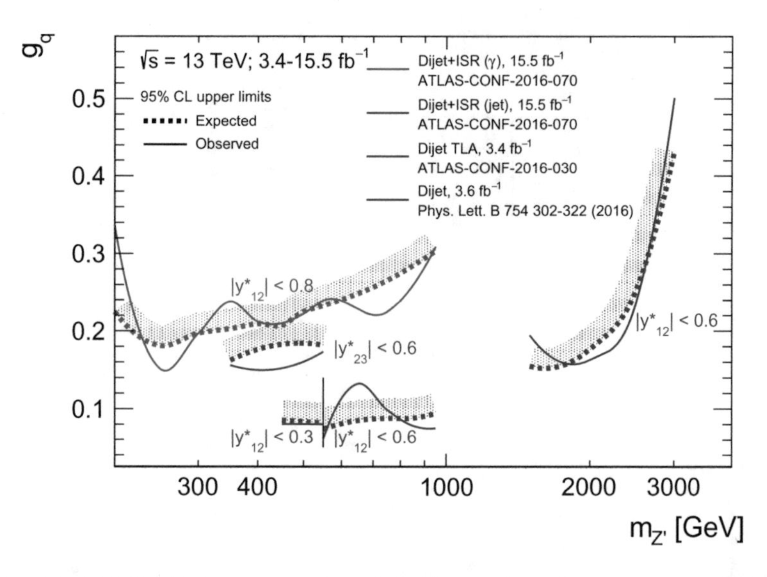

Fig. 7.11 The 95% C.L. limits on the coupling of the Z' to quarks g_q, is shown versus the Z' mass. The solid curves show the observed upper limit, and the dashed curves show the expected upper limit. The limits from the high mass dijet analysis (dark blue curve), dijet + γ analysis (red curve), dijet + jet analysis (purple curve), and the Trigger-object Level Analysis (TLA) (blue curve) are presented. Figure adapted from [32]

In order to produce this plot, the limits on $\sigma \times$ A $\times$ BR for the Z' model are used to calculate the ratio between the observed limit and the theoretical prediction, $\frac{\sigma_{\text{limit}}}{\sigma_{\text{theory}}}$, for each mass and coupling point, as shown in Fig. 7.8 for the high mass dijet analysis. Using the fact that the signal cross-section scales proportionally to g_q^2, and using the lowest coupling value which was excluded for a particular mass point as a reference $g_{q_{\text{ref}}}$, the value of the limit on the coupling $g_{q_{\text{limit}}}$ can be calculated. The limit is calculated as

$$g_{q_{\text{limit}}} = \sqrt{g_{q_{\text{ref}}}^2 \left(\frac{\sigma_{\text{limit}}}{\sigma_{\text{theory}} \beta} \right)}, \tag{7.7}$$

where β is a scale factor equal to 1 if the theoretical cross-section includes the decay of the Z' to beauty quarks, and equal to 1.25 if the decays to beauty quarks are not included. The Z' signal samples used in the high mass dijet analysis and the TLA did not include the decays of beauty quarks. Note that when determining $g_{q_{\text{ref}}}$, exclusion means that the ratio $\frac{\sigma_{\text{limit}}}{\sigma_{\text{theory}} \beta} < 1$. As an example, consider the 1.5 TeV mass point in the high mass dijet analysis. Using Fig. 7.8, and dividing the $\frac{\sigma_{\text{limit}}}{\sigma_{\text{theory}}}$ values by $\beta = 1.25$, we see that the lowest coupling value which was excluded ($\frac{\sigma_{\text{limit}}}{\sigma_{\text{theory}} \beta} < 1$) is $g_{q_{\text{ref}}} = 0.2$, and that $\frac{\sigma_{\text{limit}}}{\sigma_{\text{theory}} \beta}$ is equal to 1.2/1.25 = 0.96. Substituting these values into Eq. (7.7), we obtain $g_{q_{\text{limit}}} = 0.196$. This value is displayed in Fig. 7.11.

Our limits in the plane of the coupling to quarks versus Z' mass can also be compared to existing limits on the baryonic Z' model, by scaling our limits up by a

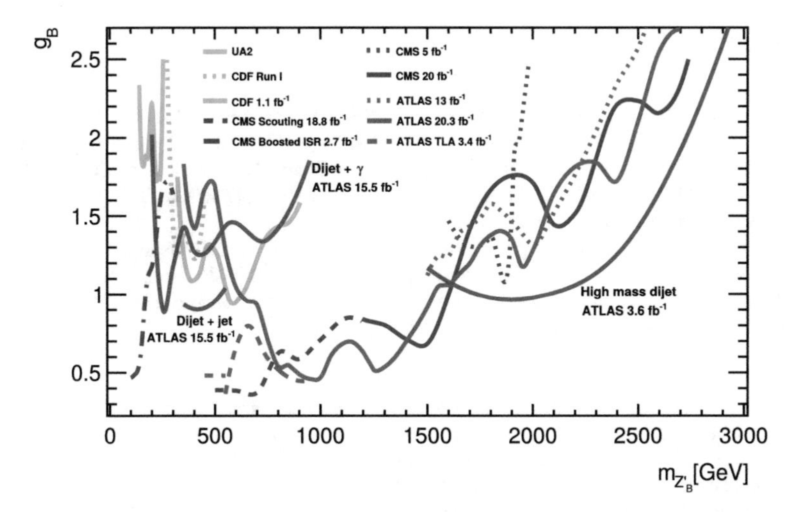

Fig. 7.12 The 95% C.L. limits on the coupling of the baryonic Z' to quarks g_B, versus the baryonic Z' mass $m_{Z'_B}$, are shown from a variety of different experiments (UA2, CDF, CMS and ATLAS). The labelled curves shown in red are the observed limits obtained from the analyses described in this thesis. This plot is adapted from [33, 34]

factor of 6 and overlaying them. The factor of 6 is due to the difference in the definition of the coupling to quarks between the two models, as described in Sect. 5.1.2. The resulting plot is shown in Fig. 7.12, where the analyses described in this thesis are shown by red labelled curves. Two separate plots showing the limits in the mass region below 1 TeV, and the limits in the mass region above 1 TeV, are shown in Appendix G.

Figure 7.12[1] shows that the high mass dijet analysis sets stronger limits than all the previous dijet analyses in the mass region above ∼1.6 TeV, and it extends the mass reach of the limits to ∼2.9 TeV. The limits from the dijet + ISR analyses provide the most stringent limits across the majority of the mass region between ∼220 and 450 GeV (with the exception of the mass region ∼315–350 GeV). This proves that the technique used in the dijet + ISR analyses has successfully enabled us to target the region below 500 GeV, and has allowed us to exceed the limits set by some of the older experiments.

Note that a CMS analysis [35] using events in which a potential resonance is boosted by initial state radiation, and is reconstructed as a single large-radius jet was released in July 2016, and is also displayed on the plot. The release of this result was after the first dijet + ISR result by ATLAS in June 2016, and before the second dijet + ISR result by ATLAS in August 2016 displayed on this plot.

[1] A description of the extraction for all the limits was provided in the caption of Fig. 2.10, with the exception of the extraction of the CMS Boosted ISR result, which was added by digitising the limit contour in [35] using WebPlotDigitizer [36].

References

1. Bolstad W (2013) Introduction to bayesian statistics. Wiley. ISBN: 9781118619216
2. Bayes M, Price M (1763) An essay towards solving a problem in the doctrine of chances. By the Late Rev. Mr. Bayes FRS, Communicated by Mr. Price, in a Letter to John Canton AMFRS
3. Particle Data Group, Olive et al KA (2014) Review of particle physics. Chin Phys C38:090001. https://doi.org/10.1088/1674-1137/38/9/090001
4. Caldwell A, Kollar D, Kröninger K (2009) BAT - The Bayesian analysis toolkit. Comput Phys Commun 180.11:2197–2209. https://doi.org/10.1016/j.cpc.2009.06.026, ISSN: 0010-4655
5. Pachal K (2015) Search for new physics in the dijet invariant mass spectrum at 8 TeV. CERN-THESIS-2015-179. Ph.D. thesis. The University of Oxford
6. The Pennsylvania State University, Likelihood & LogLikelihood. https://onlinecourses.science.psu.edu/stat504/node/27
7. Huber C (2016) Introduction to Bayesian statistics. http://blog.stata.com/2016/11/01/introduction-to-bayesian-statistics-part-1-the-basic-concepts
8. James F (2015) Lecture notes from the Terascale Statistics School. Hamburg. https://indico.desy.de/conferenceDisplay.py?confId=11244
9. Stanford J, Vardeman S (1994) Statistical methods for physical science. Methods of experimental physics. Elsevier Science. https://www.elsevier.com/books/statistical-methods-for-physical-science/stanford/978-0-12-475973-2. ISBN: 9780080860169
10. Cowan G (1998) Statistical data analysis. Oxford science publications, Clarendon Press, Oxford. ISBN: 9780198501565
11. ROOT 6.11/01 (2017) TMinuit class reference. https://root.cern.ch/doc/master/classTMinuit.html
12. Oreglia M, A study of the reactions $\psi' \rightarrow \gamma\gamma\psi$. SLAC-236. Ph.D. thesis. SLAC
13. Pöttgen R (2016) Search for dark matter with ATLAS: using events with a highly energetic jet and missing transverse momentum in proton-proton collisions at $\sqrt{s} = 8\,TeV$. Springer theses. Springer International Publishing, Berlin. http://www.springer.com/gb/book/9783319410449. ISBN: 9783319410456
14. ATLAS Collaboration (2015) Search for new phenomena in the dijet mass distribution using pp collision data at $\sqrt{s} = 8$ TeV with the ATLAS detector. Phys Rev D 91:052007. https://doi.org/10.1103/PhysRevD.91.052007. arXiv:1407.1376 [hep-ex]
15. ATLAS Collaboration (2013) Jet energy measurement with the ATLAS detector in proton-proton collisions at $\sqrt{s} = 7$ TeV. Eur Phys J C73.3:2304. https://doi.org/10.1140/epjc/s10052-013-2304-2. arXiv: 1112.6426 [hep-ex]
16. ATLAS Collaboration (2016) Luminosity determination in pp collisions at $\sqrt{s} = 8$ TeV using the ATLAS detector at the LHC. Eur Phys J C 76.12:653. https://doi.org/10.1140/epjc/s10052-016-4466-1. ISSN: 1434-6052
17. NNPDF Developers, Neural network parton distribution functions. http://nnpdf.hepforge.org
18. ATLAS Collaboration, Recommendation for using PDFs. https://twiki.cern.ch/twiki/bin/viewauth/AtlasProtected/PdfRecommendations
19. Buckley A et al (2015) LHAPDF6: parton density access in the LHC precision era. Eur Phys J C75:132. https://doi.org/10.1140/epjc/s10052-015-3318-8. arXiv:1412.7420 [hep-ph]
20. MSTW Collaboration, Martin-Stirling-Thorne-Watt parton distribution functions. http://mstwpdf.hepforge.org/
21. ATLAS Collaboration (2016) Electron and photon energy calibration with the ATLAS detector using data collected in 2015 at $\sqrt{s} = 13$ TeV. Technical report, ATLAS-PHYS-PUB-2016-015. Geneva: CERN
22. ATLAS Collaboration (2014) Electron and photon energy calibration with the ATLAS detector using LHC Run 1 data. Eur Phys J C 74.10:3071. https://doi.org/10.1140/epjc/s10052-014-3071-4, ISSN: 1434-6052
23. Metropolis N et al (1953) Equation of state calculations by fast computing machines. J Chem Phys 21.6:1087–1092. https://doi.org/10.1063/1.1699114

24. Gelfand AE, Smith AFM (1990) Sampling-based approaches to calculating marginal densities. J Am Stat Assoc 85.410:398–409. http://www.jstor.org/stable/2289776, ISSN: 01621459
25. Bayesian Analysis Toolkit Developers (2017) Bayesian analysis toolkit manual. https://github.com/bat/bat/tree/manual/doc/manual
26. Beaujean F et al (2011) Bayesian analysis toolkit in searches. https://cds.cern.ch/record/2203253
27. Markov AA (1906) Extension of the law of large numbers to dependent quantities (in Russian). Izvestiia Fiz.-Matem. Obsc h. Kazan Univ. (2nd Ser.) 15(1906):135–156
28. Hastings WK (1970) Monte Carlo sampling methods using markov chains and their applications. Biometrika 57.1:97–109. http://www.jstor.org/stable/2334940, ISSN: 00063444
29. Brooks S et al (2011) Handbook of Markov chain Monte Carlo, Handbooks of modern statistical methods. Chapman & Hall/CRC, CRC Press, Boca Raton. ISBN: 9781420079425
30. Holder M (2017) MCMC notes. http://phylo.bio.ku.edu/slides/2011_lhm_bayesian_mcmc_1.pdf
31. ATLAS Collaboration (2016) Search for light dijet resonances with the ATLAS detector using a trigger-level analysis in LHC pp collisions at $\sqrt{s} = 13$ TeV. Technical report, ATLASCONF-2016-030. Geneva: CERN
32. ATLAS Collaboration, Dark matter summary plot. https://atlas.web.cern.ch/Atlas/GROUPS/PHYSICS/CombinedSummaryPlots/EXOTICS/ATLAS_DarkMatterCoupling_Summary/history.html
33. ATLAS Collaboration, Baryonic Z' summary plot. https://atlas.web.cern.ch/Atlas/GROUPS/PHYSICS/PAPERS/EXOT-2013-11/figaux_10.png
34. Boveia A (2017) Private communication
35. CMS Collaboration (2016) Search for light vector resonances decaying to quarks at $\sqrt{s} =$ 13 TeV. Technical report CMS-PAS-EXO-16-030. Geneva: CERN
36. Rohatgi A (2017) WebPlotDigitizer - web based plot digitizer version 3.11. http://arohatgi.info/WebPlotDigitizer

Chapter 8
Conclusion and Outlook

In 2015 and 2016, the LHC delivered proton-proton collisions with an unprecedented centre-of-mass energy of $\sqrt{s} = 13\,\text{TeV}$. In this thesis results are shown from the analysis of the high energy collision data recorded by the ATLAS detector.

One of the first analyses performed using the $\sqrt{s} = 13$ TeV collision data, collected in 2015 (3.6 fb^{-1}), was the search for high mass resonances in the dijet final state. In this analysis dijet events with invariant masses ranging from 1.1 to 6.9 TeV were studied. An event display of the highest mass dijet event utilised in the search is shown in Fig. 8.1. For reference, the highest mass achieved in the previous ATLAS search was 4.5 TeV [1], demonstrating that a new high mass region of phase space was explored by the high mass dijet analysis presented in this thesis.

In the high mass dijet analysis a spectrum of dijet invariant masses was produced, and localised excesses of events above the background estimation were searched for. The largest excess observed was between approximately 1.5–1.6 GeV, with a global p-value of 0.67. This indicates that no significant excess was present in the data, and hence, there is no evidence for the presence of dijet resonances in the explored mass range.

The data are used to set stringent limits on several models of new physics, in addition to model-independent Gaussian resonance shapes. The 95% C.L. limits set on the mass of quantum black holes reached 8.3 TeV, increasing the limit achieved in Run I by more than 2 TeV. Excited quarks with masses below 5.2 TeV were excluded at 95% C.L, and heavy W' bosons were excluded below 2.6 TeV. Limits were also set on a lepto-phobic Z' dark matter mediator model for the first time. The limits in the plane of the coupling to quarks versus the mass of the Z' were used to calculate the corresponding limits on the baryonic Z' model, providing a direct comparison to the existing results. The comparison showed that our results exceeded all the existing results above 1.6 TeV, and extended the exclusion to Z' masses of up to 2.9 TeV.

The other two analyses presented in this thesis utilised $\sqrt{s} = 13$ TeV data, collected in 2015 and 2016 (15.5 fb^{-1}), to search for low mass dijet resonances. A new technique was used to overcome the trigger limitations in the low mass region.

L. A. Beresford, *Searches for Dijet Resonances*, Springer Theses,
https://doi.org/10.1007/978-3-319-97520-7_8

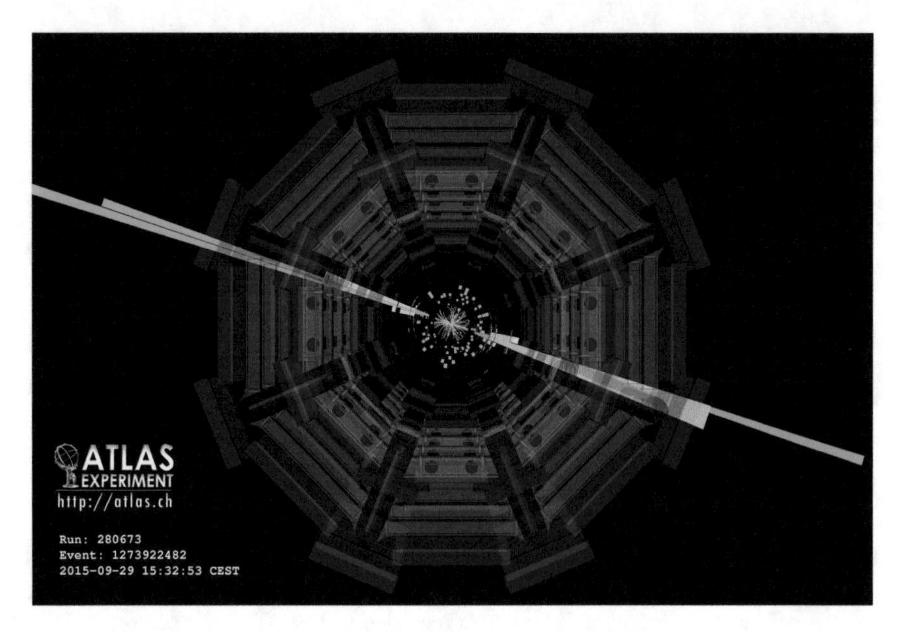

Fig. 8.1 The highest mass dijet event utilised in high mass dijet resonance search, taken from [2]. The pair of jets have a combined invariant mass of 6.9 TeV

Events in which a dijet was balanced against a high momentum photon or jet (arising from initial state radiation) were selected by triggering on the high momentum object; allowing us to efficiently gather low mass dijet events. Dijet masses in the range 200–1500 GeV were investigated in the dijet $+\gamma$ analysis and masses in the range 300–600 GeV were investigated in the dijet +jet analysis. No evidence for resonances was observed in either case. Limits were placed on model-independent Gaussian resonance shapes and on a lepto-phobic Z' dark matter mediator model. The corresponding limits set on a baryonic Z' model showed that our results exceed all the previous results in the mass range between 220–315 GeV and 350–450 GeV, surpassing existing limits from the UA2, CDF, CMS and ATLAS experiments.

Since the release of the results contained in this thesis, the high mass dijet analysis has been performed with the full dataset collected in 2015 and 2016 (37 fb^{-1} of data) [3], further driving down the limits in the high mass region, as shown in Fig. 8.2. The statistical package which I worked on during my DPhil was utilised in the production of these results. Additionally, CMS has performed a high mass dijet analysis using 12.9 fb^{-1} of data [4].

In the low mass region, CMS has released an updated version of their ISR + large-radius jet analysis with 36 fb^{-1} of data [6]. Figure 8.3 shows a comparison between the 95% C.L. limits set by this analysis and the 95% C.L. limits set by other low mass dijet analyses, including the results from the dijet + ISR analyses described in this thesis. The limits shown are for the Z' dark matter mediator model in the plane of the Z' coupling to quarks g_q, versus the mass of the Z'. The comparison

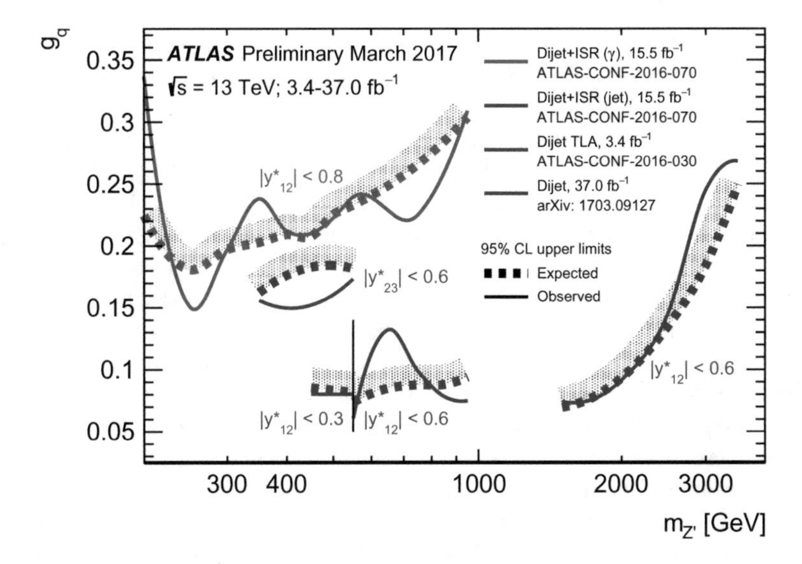

Fig. 8.2 The 95% C.L. limits on the coupling of the Z' to quarks g_q, is shown versus the Z' mass. The solid curves show the observed upper limit, and the dashed curves show the expected upper limit. The limits from the latest high mass dijet analysis using 37 fb^{-1} of data is shown by the dark blue curve. The curves for the other analyses are the same as those displayed previously. Figure taken from [5]

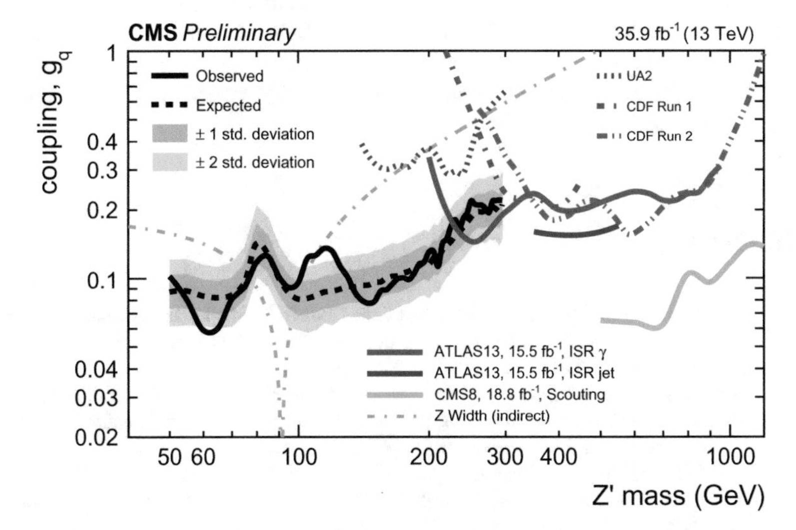

Fig. 8.3 The 95% C.L. limits on the coupling of the Z' to quarks g_q versus the mass of the Z' are shown from a variety of different experiments (UA2, CDF, CMS and ATLAS). The results from the dijet + ISR analyses described in this thesis are shown by the red solid line (dijet + γ) and by the blue solid line (dijet + jet). The latest CMS results, shown in black, supersede the ATLAS dijet + ISR results below 225 GeV and can probe masses down to 50 GeV. Figure taken from [6]

shows that these new results exceed those obtained by the dijet + ISR analyses in this thesis below 225 GeV, and that this analysis can set limits on Z' masses down to 50 GeV. This is achieved by utilising a large-radius jet, rather than a resolved pair of jets. This technique relies on the fact that in order to balance the momentum of the high p_T ISR object, a very light resonance would receive a large Lorentz boost. This causes the decay products of the resonance to merge, such that they cannot be resolved as two distinct jets, and are instead reconstructed as a large-radius jet. The result released by CMS shows a small excess at 115 GeV (2.9σ local significance, 2.2σ global significance). An ATLAS analysis in the ISR + large-radius jet channel is underway, and the results from this will be very interesting to see.

CMS has also released a result in which b-tagging is applied to the large-radius jet in the ISR + large-radius jet channel [7], using 36 fb^{-1} of data. This analysis observed the decay of a Z boson to a $b\bar{b}$ pair with a local significance of 5.1σ, and observed an excess at the Higgs mass with a local significance of 1.5σ. This indicates that the techniques applied here can successfully identify low mass resonances in the di-b-jet final state, which is significant for studying Standard Model processes, and for searching for new particles at low di-b-jet mass. The resolved and boosted ISR analyses could, in principle, be performed using trigger jets, recorded with reduced information, as utilised in the ATLAS Trigger-object Level analysis [8]. This would further increase the data yield obtained at low dijet masses, enhancing the sensitivity to new particles with a low production cross-section.

With an expected data yield of 150 fb^{-1} by 2019, and new techniques at our disposal, there is much hope for finding evidence for Beyond Standard Model particles at the LHC. The dijet final state remains to be a promising place to search for new particles, and the quest to understand the particles and forces which make up our universe continues.

References

1. ATLAS Collaboration (2015) Search for new phenomena in the dijet mass distribution using pp collision data at $\sqrt{s} = 8$ TeV with the ATLAS detector. Phys Rev D 91:052007. https://doi.org/10.1103/PhysRevD.91.052007, arXiv:1407.1376 [hep-ex]
2. ATLAS Collaboration, Auxiliary Material for Search for new phenomena in dijet mass and angular distributions from pp collisions at $\sqrt{s} = 13$ TeV with the ATLAS detector. https://atlas.web.cern.ch/Atlas/GROUPS/PHYSICS/PAPERS/EXOT-2015-02/
3. ATLAS Collaboration (2017) Search for new phenomena in dijet events using 37 fb^{-1} of pp collision data collected at ps = 13 TeV with the ATLAS detector. arXiv: 1703.09127 [hep-ex]
4. CMS Collaboration (2017) Search for dijet resonances in proton-proton collisions at $\sqrt{s} = 13$ TeV and constraints on dark matter and other models. Phys Lett B 769:520–542. https://doi.org/10.1016/j.physletb.2017.02.012, ISSN: 0370-2693
5. ATLAS Collaboration, Dark matter summary plot. https://atlas.web.cern.ch/Atlas/GROUPS/PHYSICS/CombinedSummaryPlots/EXOTICS/ATLAS_DarkMatterCoupling_Summary/history.html
6. CMS Collaboration (2017) Search for light vector resonances decaying to a quark pair produced in association with a jet in proton-proton collisions at $\sqrt{s} = 13$ TeV. Technical report, CMS-PAS-EXO-17-001. Geneva: CERN

7. CMS Collaboration (2017) Inclusive search for the standard model Higgs boson produced in pp collisions at $\sqrt{s} = 13\,\text{TeV}$ using $H \to b\bar{b}$ decays. Technical report, CMS-PAS-HIG-17-010. Geneva: CERN
8. ATLAS Collaboration (2016) Search for light dijet resonances with the ATLAS detector using a trigger-level analysis in LHC pp collisions at $\sqrt{s} = 13\,\text{TeV}$. Technical report, ATLASCONF-2016-030. Geneva: CERN

Appendix A
Jet Cleaning

A jet is identified as *BadLoose*, i.e. likely to be a fake jet if it satisfies any of the following criteria:

1. $f_{HEC} > 0.5$ and $|f_Q^{HEC}| > 0.5$ and $\langle Q \rangle > 0.8$;
2. $|E_{neg}| > 60$ GeV;
3. $f_{EM} > 0.95$ and $f_Q^{LAr} > 0.8$ and $\langle Q \rangle > 0.8$ and $|\eta| < 2.8$;
4. $f_{max} > 0.99$ and $|\eta| < 2$;
5. $f_{EM} < 0.05$ and $f_{ch} < 0.05$ and $|\eta| < 2$;
6. $f_{EM} < 0.05$ and $|\eta| \geq 2$.

L. A. Beresford, *Searches for Dijet Resonances*, Springer Theses,
https://doi.org/10.1007/978-3-319-97520-7

Appendix B
Mass Spectra Binning

The final derived bin edges for the high mass dijet analysis, in GeV:

946, 976, 1006, 1037, 1068, 1100, 1133, 1166, 1200, 1234, 1269, 1305, 1341, 1378, 1416, 1454, 1493, 1533, 1573, 1614, 1656, 1698, 1741, 1785, 1830, 1875, 1921, 1968, 2016, 2065, 2114, 2164, 2215, 2267, 2320, 2374, 2429, 2485, 2542, 2600, 2659, 2719, 2780, 2842, 2905, 2969, 3034, 3100, 3167, 3235, 3305, 3376, 3448, 3521, 3596, 3672, 3749, 3827, 3907, 3988, 4070, 4154, 4239, 4326, 4414, 4504, 4595, 4688, 4782, 4878, 4975, 5074, 5175, 5277, 5381, 5487, 5595, 5705, 5817, 5931, 6047, 6165, 6285, 6407, 6531, 6658, 6787, 6918, 7052, 7188, 7326, 7467, 7610, 7756, 7904, 8055, 8208, 8364, 8523, 8685, 8850, 9019, 9191, 9366, 9544, 9726, 9911, 10100, 10292, 10488, 10688, 10892, 11100, 11312, 11528, 11748, 11972, 12200, 12432, 12669, 12910, 13156

The final derived bin edges for the dijet + ISR analyses, in GeV:

169, 180, 191, 203, 216, 229, 243, 257, 272, 287, 303, 319, 335, 352, 369, 387, 405, 424, 443, 462, 482, 502, 523, 544, 566, 588, 611, 634, 657, 681, 705, 730, 755, 781, 807, 834, 861, 889, 917, followed by the high mass dijet analysis bin edges given above.

L. A. Beresford, *Searches for Dijet Resonances*, Springer Theses,
https://doi.org/10.1007/978-3-319-97520-7

Appendix C
Event Yields

See Tables C.1, C.2 and C.3.

Table C.1 Event yields for the full 3.6 fb^{-1} used in the high mass dijet analysis. The Trigger OR includes events passing any of the following triggers: L1_J75, L1_J100, HLT_j360, HLT_3j175 or HLT_4j85

Selection criteria	N_{events}
All	35477718
Event quality	35393381
Primary vertex requirement	35391453
Trigger OR	23350594
At least two jets with $p_T > 50$ GeV	23020926
Leading jet $p_T > 200$ GeV	12740838
HLT_j360 trigger	11995952
Jet cleaning	11988448
Leading jet $p_T > 440$ GeV	4979860
$m_{jj} > 1100$ GeV	2480182
$\lvert y^*_{12}\rvert < 0.6$	677852

L. A. Beresford, *Searches for Dijet Resonances*, Springer Theses,
https://doi.org/10.1007/978-3-319-97520-7

Table C.2 Event yields for the full 15.5 fb^{-1} used in the dijet + γ analysis. The pre-selection includes the trigger, event quality and primary vertex requirements, in addition to requiring at least two jets with $p_T > 25$ GeV and within $|\eta| < 2.8$, and one tight ID isolated photon with $p_T > 10$ GeV

Selection	N_{events}
Pre-selection	3097173
Jet cleaning	3094508
Photon $p_T > 150\,\text{GeV}$	1346530
$\lvert y^*_{12}\rvert < 0.8$	903526
$\Delta R_{\text{ISR,close-jet}} > 0.85$	854666
$m_{jj} > 160\,\text{GeV}$	198009

Table C.3 Event yields for the full 15.5 fb^{-1} used in the dijet + jet analysis. The pre-selection includes the trigger, event quality, and primary vertex requirements, in addition to requiring at least two jets with $p_T > 25$ GeV and within $|\eta| < 2.8$

Selection	N_{events}
Pre-selection	20008992
Jet cleaning	19980876
Three Jets with $p_T > 25\,\text{GeV}$	17309262
Leading Jet $p_T > 430\,\text{GeV}$	9319099
$\lvert y^*_{23}\rvert < 0.6$	5708908
$m_{jj} > 270\,\text{GeV}$	1507667

Appendix D
Additional Search Phase Plots

See Figs. D.1 and D.2.

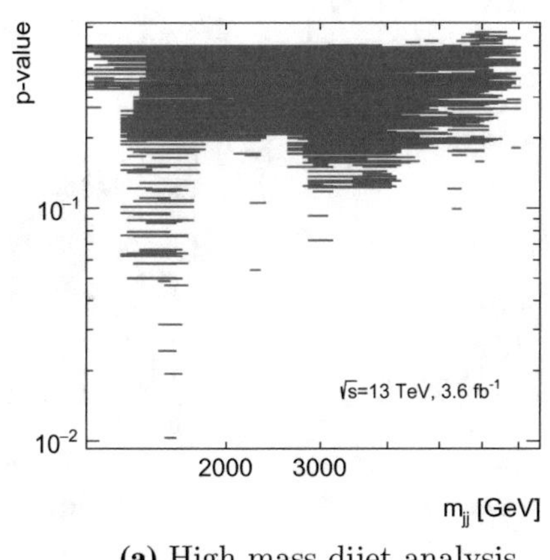

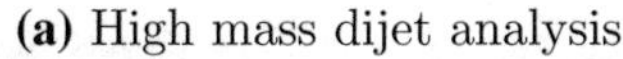

(a) High mass dijet analysis

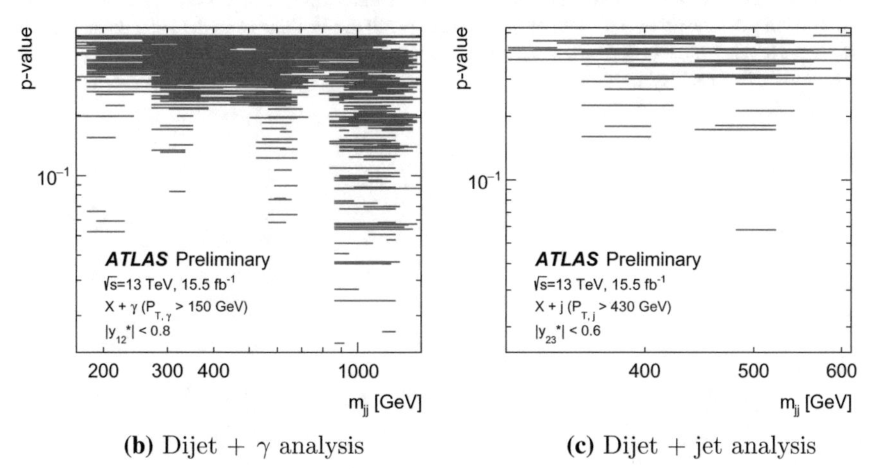

(b) Dijet + γ analysis

(c) Dijet + jet analysis

Fig. D.1 These figures show the local p-value in each mass window considered in the search

L. A. Beresford, *Searches for Dijet Resonances*, Springer Theses,
https://doi.org/10.1007/978-3-319-97520-7

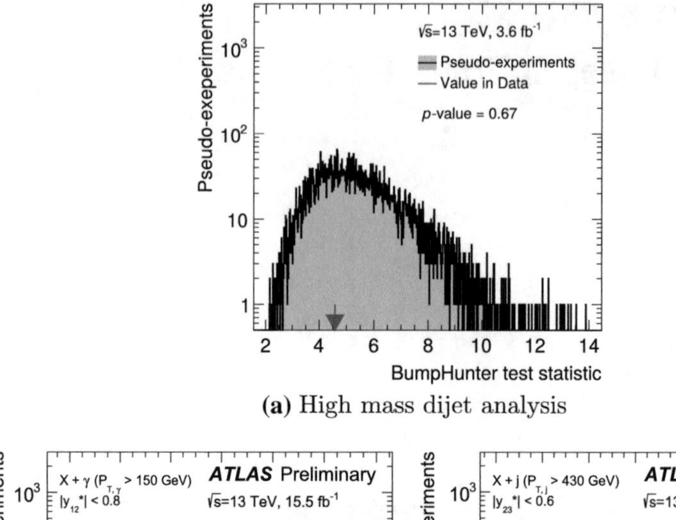

(a) High mass dijet analysis

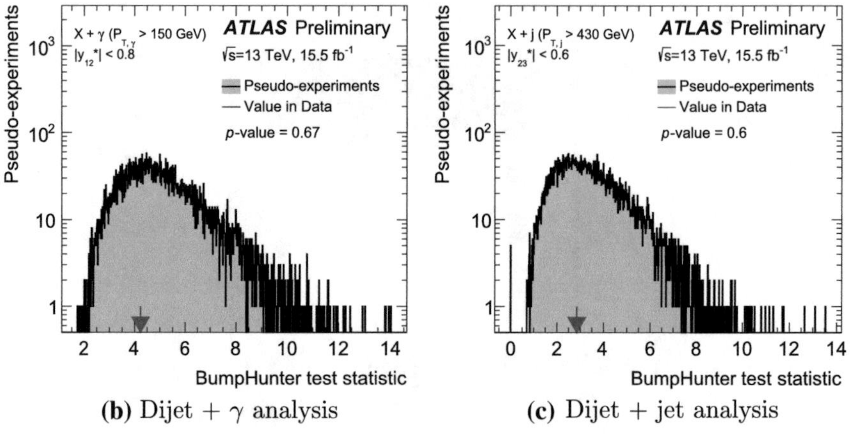

(b) Dijet + γ analysis

(c) Dijet + jet analysis

Fig. D.2 These figures show the distribution of BUMPHUNTER test statisics from pseudo-experiments, and the observed value of the BUMPHUNTER test statistic from data is indicated by the arrow. A global p-value is derived by calculating the fraction of pseudo-experiments with a p-value greater than the observed p-value

Appendix E
Additional Limit Setting Plots

See Figs. E.1, E.2 and E.3.

L. A. Beresford, *Searches for Dijet Resonances*, Springer Theses,
https://doi.org/10.1007/978-3-319-97520-7

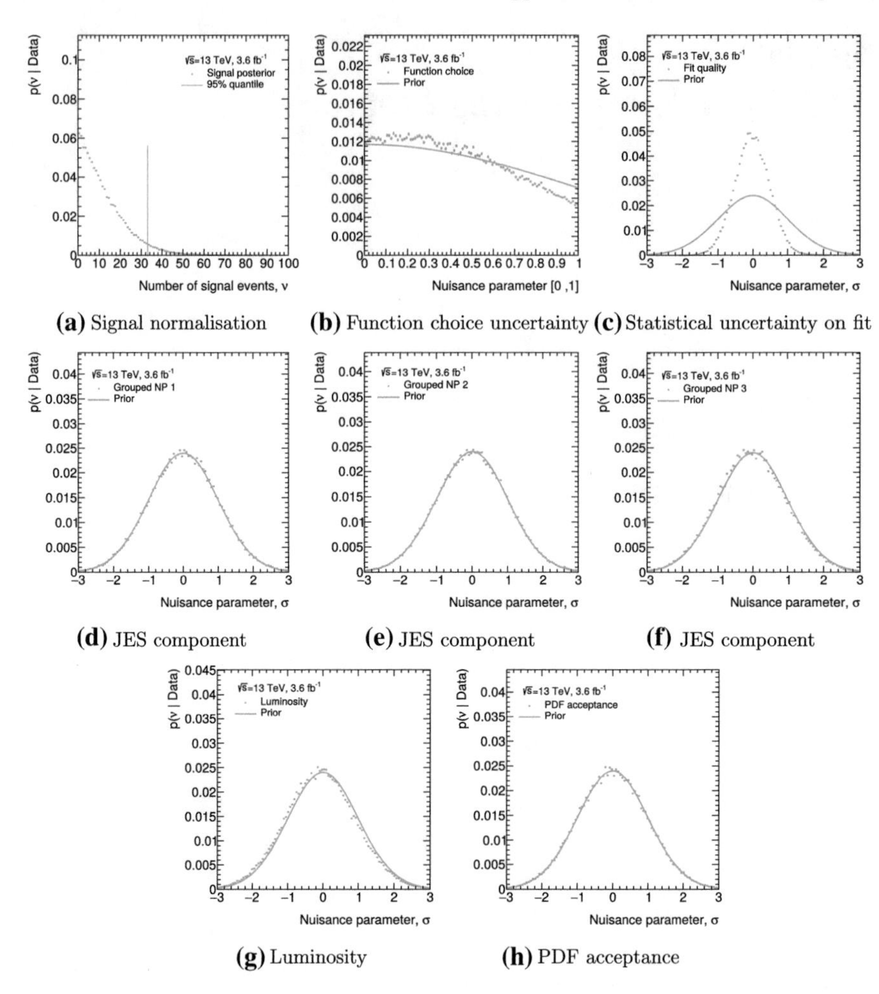

(a) Signal normalisation **(b)** Function choice uncertainty **(c)** Statistical uncertainty on fit

(d) JES component **(e)** JES component **(f)** JES component

(g) Luminosity **(h)** PDF acceptance

Fig. E.1 For a 5 TeV q* in the high mass dijet analysis, a comparison between the prior (solid line) and the marginalised posterior (dotted line) is displayed. In Figure **a** the solid line indicates the 95% quantile of the number of signal events ν, not the prior

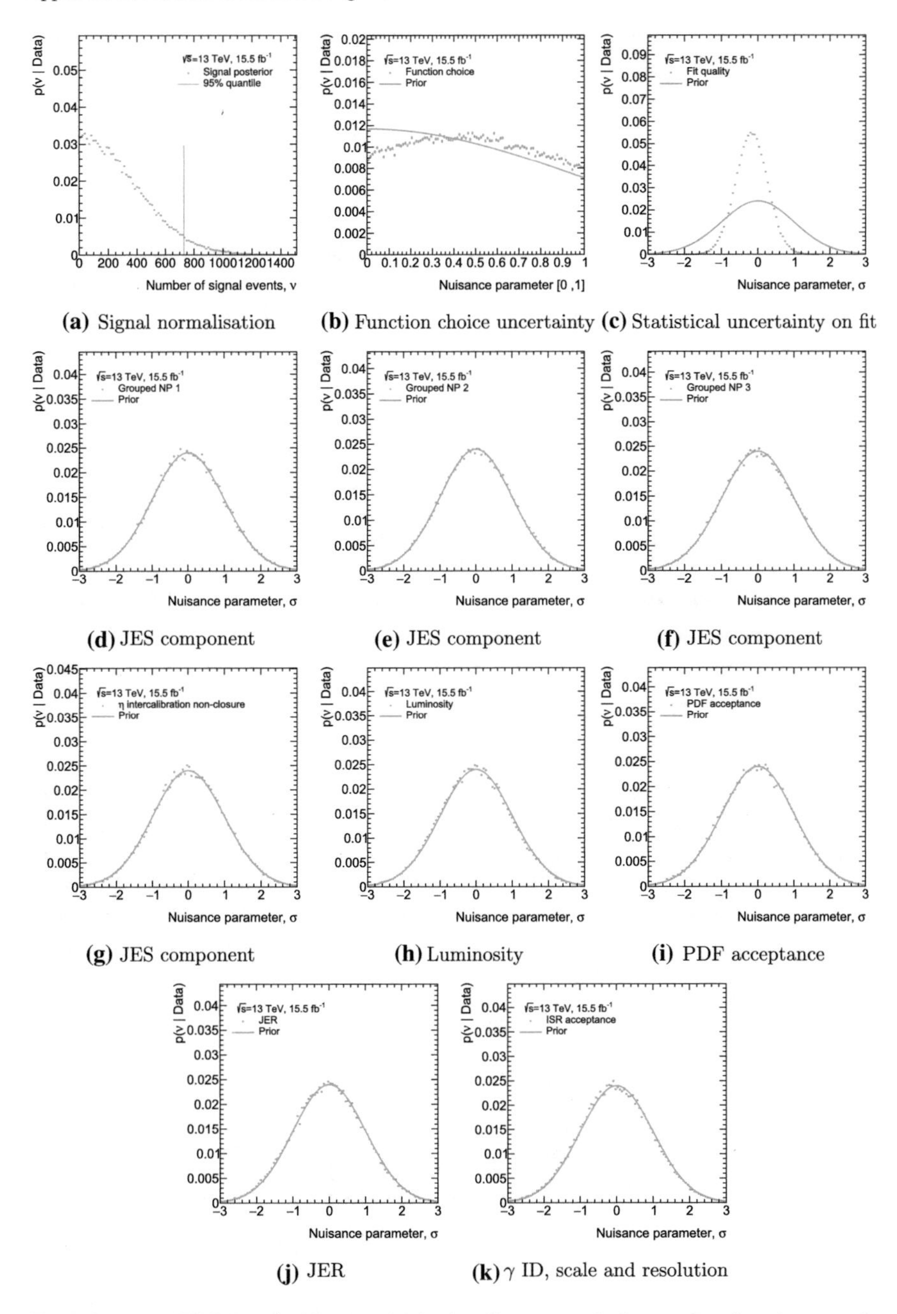

(a) Signal normalisation **(b)** Function choice uncertainty **(c)** Statistical uncertainty on fit

(d) JES component **(e)** JES component **(f)** JES component

(g) JES component **(h)** Luminosity **(i)** PDF acceptance

(j) JER **(k)** γ ID, scale and resolution

Fig. E.2 For a 450 GeV Z' with $g_q = 0.3$ in the dijet + γ analysis, a comparison between the prior (solid line) and the marginalised posterior (dotted line) is displayed. In Figure **a** the signal normalisation ν (solid line indicates the 95% quantile, not the prior)

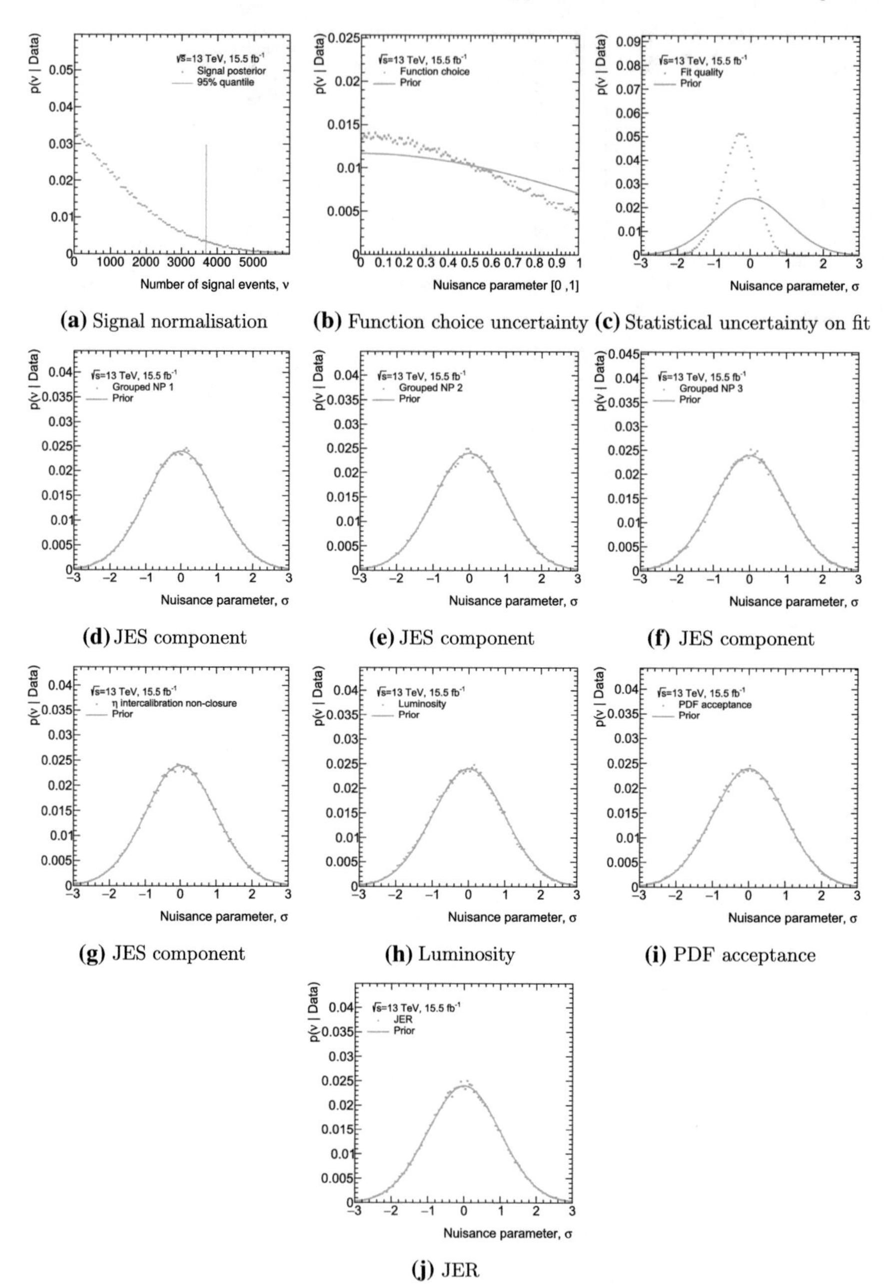

(a) Signal normalisation **(b)** Function choice uncertainty **(c)** Statistical uncertainty on fit

(d) JES component **(e)** JES component **(f)** JES component

(g) JES component **(h)** Luminosity **(i)** PDF acceptance

(j) JER

Fig. E.3 For a 450 GeV Z' with $g_q = 0.3$ in the dijet + jet analysis, a comparison between the prior (solid line) and the marginalised posterior (dotted line) is displayed. In Figure **a** the solid line indicates the 95% quantile of the number of signal events ν, not the prior

Appendix F
Gaussian Limit Tables

See Tables F.1, F.2 and F.3.

Table F.1 This table displays the dijet $+\gamma$ analysis upper limits set on $\sigma \times A \times BR$ at 95% C.L. for Gaussian signal shapes with mean mass m_G, for various width-to-mass ratios. The smallest width-to-mass ratio (Res.) corresponds to the detector mass resolution. Each of the limits have been corrected for experimental inefficiencies through division by 0.81, the average photon efficiency

m_G [GeV]	Limit [pb] (σ_G/m_G = Res.)	Limit [pb] (σ_G/m_G = 7%.)	Limit [pb] (σ_G/m_G = 10%)
200	0.098	0.16	–
250	0.025	0.033	0.045
300	0.05	0.072	0.088
350	0.035	0.051	0.066
400	0.026	0.034	0.044
450	0.023	0.031	0.038
500	0.019	0.027	0.035
550	0.019	0.032	0.044
600	0.027	0.038	0.043
650	0.021	0.026	0.025
700	0.0089	0.013	0.015
750	0.0073	0.0092	0.011
800	0.0063	0.0088	0.012
900	0.014	0.015	0.016
1000	0.0077	0.013	0.018
1200	0.0096	0.015	0.02

L. A. Beresford, *Searches for Dijet Resonances*, Springer Theses,
https://doi.org/10.1007/978-3-319-97520-7

Table F.2 This table displays the dijet + jet analysis upper limits set on $\sigma \times A \times BR$ at 95% C.L. for Gaussian signal shapes with mean mass m_G, for various width-to-mass ratios. The smallest width-to-mass ratio (Res.) corresponds to the detector mass resolution

m_G [GeV]	Limit [pb] (σ_G/m_G = Res.)	Limit [pb] (σ_G/m_G = 7%.)	Limit [pb] (σ_G/m_G = 10%)
350	0.12	–	–
400	0.16	0.19	0.22
450	0.066	0.11	0.14
500	0.13	0.18	0.22
550	0.052	–	–

Table F.3 This table displays the high mass dijet analysis upper limits set on $\sigma \times A \times BR$ at 95% C.L. for Gaussian signal shapes with mean mass m_G, for various width-to-mass ratios. The smallest width-to-mass ratio (Res.) corresponds to the detector mass resolution

m_G [GeV]	Limit [pb] (σ_G/m_G = Res.)	Limit [pb] (σ_G/m_G = 7%.)	Limit [pb] (σ_G/m_G = 10%)	Limit [pb] (σ_G/m_G = 15%)
1200	0.28	–	–	–
1250	0.16	–	–	–
1300	0.25	0.4	–	–
1350	0.31	0.52	-	–
1400	0.3	0.64	1.0	–
1450	0.22	0.67	0.97	–
1500	0.26	0.64	0.79	–
1550	0.31	0.54	0.52	–
1600	0.29	0.36	0.29	0.28
1650	0.2	0.21	0.21	0.23
1700	0.093	0.14	0.16	0.21
1750	0.056	0.1	0.13	0.18
1800	0.049	0.089	0.12	0.17
1850	0.056	0.087	0.11	0.16
1900	0.066	0.087	0.11	0.14
1950	0.069	0.091	0.11	0.14
2000	0.062	0.092	0.11	0.13
2100	0.051	0.11	0.1	0.11
2200	0.065	0.11	0.091	0.1
2300	0.068	0.084	0.078	0.097
2400	0.04	0.057	0.069	0.11
2500	0.025	0.048	0.061	0.1
2600	0.026	0.045	0.068	0.098
2700	0.026	0.051	0.067	0.091

(continued)

Table F.3 (continued)

m_G [GeV]	Limit [pb] (σ_G/m_G = Res.)	Limit [pb] ($\sigma_G/m_G = 7\%$.)	Limit [pb] ($\sigma_G/m_G = 10\%$)	Limit [pb] ($\sigma_G/m_G = 15\%$)
2800	0.025	0.053	0.064	0.084
2900	0.028	0.051	0.058	0.076
3000	0.028	0.044	0.051	0.072
3100	0.025	0.038	0.047	0.066
3200	0.021	0.032	0.042	0.059
3300	0.014	0.028	0.04	0.051
3400	0.012	0.027	0.037	0.045
3500	0.013	0.026	0.034	0.037
3600	0.014	0.025	0.031	0.031
3700	0.014	0.024	0.028	0.027
3800	0.013	0.023	0.025	0.024
3900	0.013	0.021	0.023	0.021
4000	0.013	0.019	0.02	0.018
4200	0.011	0.018	0.018	0.015
4400	0.01	0.016	0.015	0.012
4600	0.0092	0.014	0.013	0.011
4800	0.0078	0.011	0.01	0.0084
5000	0.0059	0.0089	0.0081	0.0081
5200	0.004	0.0056	0.0062	0.0067
5400	0.0033	0.0045	0.0052	0.0059
5600	0.0031	0.0041	0.0046	–
5800	0.003	0.0038	0.0043	–
6000	0.0029	0.0035	–	–
6200	0.0028	–	–	–
6400	0.0026	–	–	–
6600	0.0023	–	–	–

Appendix G
Dark Matter Summary Plots

See Figs. G.1 and G.2.

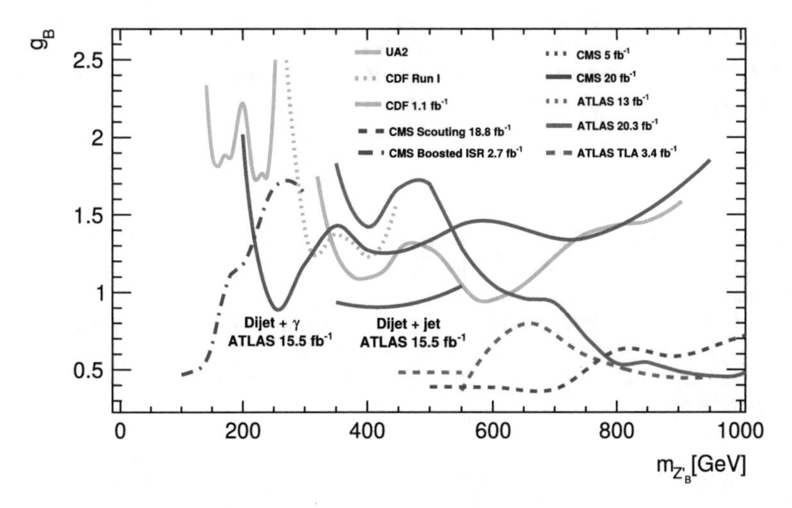

Fig. G.1 The 95% C.L. limits on the coupling of the baryonic Z' to quarks g_B, versus the mass of the Z', $m_{Z'_B}$, are shown for the mass region below 1 TeV. The limits shown are from a variety of different experiments (UA2, CDF, CMS and ATLAS). This plot is adapted from [1, 2]. The labelled curves shown in red are the observed limits obtained from the dijet + ISR analyses described in this thesis. A description of the extraction for all of the other limits was provided in the caption of Fig. 2.10, with the exception of the extraction of the CMS Boosted ISR result, which was added by digitising the limit contour in [3] using WebPlotDigitizer [4]

L. A. Beresford, *Searches for Dijet Resonances*, Springer Theses,
https://doi.org/10.1007/978-3-319-97520-7

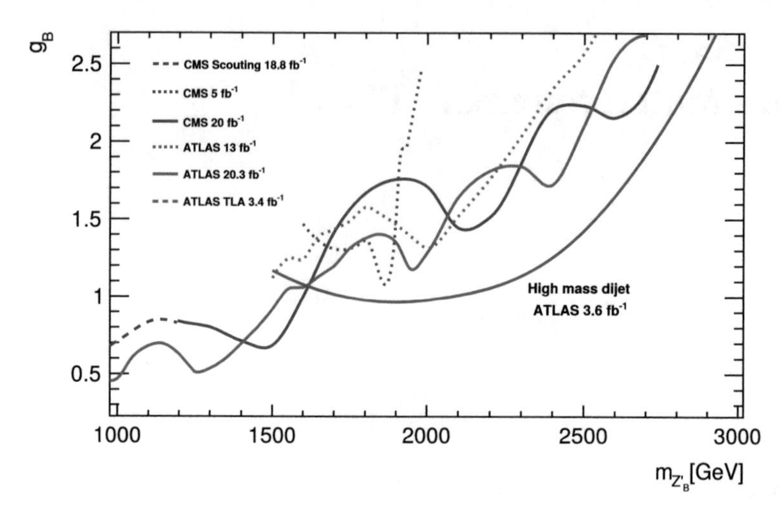

Fig. G.2 The 95% C.L. limits on the coupling of the baryonic Z' to quarks g_B, versus the mass of the Z', $m_{Z'_B}$, are shown for the mass region above 1 TeV. The limits shown are from ATLAS and CMS. This plot is adapted from [1, 2]. The labelled curves shown in red are the observed limits obtained from the high mass dijet analysis described in this thesis. A description of the extraction for all of the other limits was provided in the caption of Fig. 2.10

References

1. ATLAS Collaboration, Baryonic Z' summary plot. https://atlas.web.cern.ch/Atlas/GROUPS/PHYSICS/PAPERS/EXOT-2013-11/figaux_10.png
2. Boveia A (2017) Private communication
3. CMS Collaboration (2016) Search for light vector resonances decaying to quarks at $\sqrt{s} =$ 13 TeV. Technical report, CMS-PAS-EXO-16-030. Geneva: CERN
4. Rohatgi A (2017) WebPlotDigitizer - web based plot digitizer version 3.11. http://arohatgi.info/WebPlotDigitizer

About the Author

I am currently a Junior Research Fellow at St. John's College, University of Oxford, and a member of the ATLAS collaboration at CERN. The focus of my research is the search for exotic new particles in jet final states. I obtained my DPhil in Particle Physics from the University of Oxford under the supervision of Prof. B. Todd Huffman and Prof. Çiğdem İşsever. During my DPhil I searched for new particles which decay into pairs of jets using the first 13 TeV proton-proton collision data recorded by the ATLAS detector, and contributed to the estimation of the uncertainty on the jet energy scale for high momentum jets.

I became a member of the ATLAS collaboration in 2012 when I was a Master's student at the University of Manchester. During my Master's degree I worked on measuring the high-mass Drell–Yan differential cross-section in the muon decay channel under the supervision of Prof. Terry Wyatt.

Personal Contributions

The results presented in this thesis have been produced in collaboration between myself and other members of the ATLAS Collaboration. The focus of this thesis is on my contributions to obtaining these results; however, in order to make them

L. A. Beresford, *Searches for Dijet Resonances*, Springer Theses,
https://doi.org/10.1007/978-3-319-97520-7

understandable and to put them into context a concise account of all the stages involved is included. My main contributions are listed explicitly below.

Chapter 4: Physics Object Reconstruction in ATLAS

I produced all of the jet punch-through results shown in this chapter (with the exception of the jet-punch through correction). This includes data-Monte Carlo comparisons of variables related to jet punch-through, and the derivation of the systematic uncertainty associated with the jet punch-through correction. The majority of the jet punch-through code was written by me, and I have documented this code in [1], the other parts are based on the Run I uncertainty code [2], and the `JES_ResponseFitter` package [3].

Chapter 5: Dijet Invariant Mass Spectra

I performed signal optimisation studies for the dijet + γ analysis, contributing to the design of the analysis selection. I wrote the signal optimisation code for these studies.

Chapter 6: Searching for Resonances

I performed background estimation studies and validated the search procedure for the high mass dijet analysis and the dijet + ISR analyses using Monte Carlo. This included performing tests for the introduction of spurious signals, and for the robustness of the procedure in the presence of a signal. I produced the final search phase results for the high mass dijet analysis and for the dijet + ISR analyses using data.

Chapter 7: Limit Setting

I performed all of the limit setting for a range of new physics models and model-independent Gaussian shapes for both the high mass dijet analysis and the dijet + ISR analyses.

In order to produce the results shown in Chapters 6 and 7, I adapted and developed the Run I statistical analysis code [4], which utilises the BUMPHUNTER package [5] and the Bayesian Analysis Toolkit [6]. I have documented the code used to produce the final results in [7]. For the dijet + ISR analysis I was solely responsible for the code development, and for the high mass dijet analysis the code development was performed in collaboration with Katherine Pachal.

References

1. Beresford L (2016) Punch-through uncertainty. https://twiki.cern.ch/twiki/pub/Sandbox/PunchThroughUncertainties2015/PTUncerts.pdf
2. Gupta S (2015) A study of longitudinal hadronic shower leakage and the development of a correction for its associated effects at $\sqrt{s} = 8$ TeV with the ATLAS detector. CERN-THESIS-2015-332. Ph.D. thesis. The University of Oxford
3. ATLAS JetEtmiss performance group, JES_ResponseFitter. https://svnweb.cern.ch/trac/atlasperf/browser/CombPerf/JetETMiss/JetCalibrationTools/DeriveJES/trunk/JES_ResponseFitter
4. Pachal K (2015) Search for new physics in the dijet invariant mass spectrum at 8 TeV. CERN-THESIS-2015-179. Ph.D. thesis. The University of Oxford

5. Choudalakis G (2011) On hypothesis testing, trials factor, hypertests and the BumpHunter. In: Proceedings, PHYSTAT 2011 workshop on statistical issues related to discovery claims in search experiments and unfolding, Geneva. arXiv:1101.0390 [physics.data-an]
6. Caldwell A, Kollar D, Kröninger K (2009) BAT - The Bayesian analysis toolkit.. Comput Phys Commun 180.11:2197–2209. https://doi.org/10.1016/j.cpc.2009.06.026, ISSN: 0010-4655
7. Beresford L, Pachal K (2017) Statistical analysis tool used in ATLAS dijet resonance searches. Technical report, ATL-COM-GEN-2017-001. Geneva: CERN

Selected Publications and Conference Notes

Listed below are a selection of publications and public conference notes which I have contributed to. For [4] I was also an editor for this conference note in addition to contributing to the analysis.

1. ATLAS Collaboration (2016) Search for new phenomena in dijet mass and angular distributions from pp collisions at $\sqrt{s} = 13\,\text{TeV}$ with the ATLAS detector. Phys Lett B 754:302–322. https://doi.org/10.1016/j.physletb.2016.01.032, arXiv:1512.01530 [hep-ex]
2. ATLAS Collaboration (2017) Jet energy scale measurements and their systematic uncertain-ties in proton-proton collisions at $\sqrt{s} = 13\,\text{TeV}$ with the ATLAS detector. Phys Rev D96.7. https://doi.org/10.1103/PhysRevD.96.072002, arXiv:1703.09665 [hep-ex]
3. ATLAS Collaboration (2016) Search for new light resonances decaying to jet pairs and produced in association with a photon in proton-proton collisions at $\sqrt{s} = 13\,\text{TeV}$ with the ATLAS detector. Technical report, ATLAS-CONF-2016-029. Geneva: CERN
4. ATLAS Collaboration (2016) Search for new light resonances decaying to jet pairs and produced in association with a photon or a jet in proton-proton collisions at $\sqrt{s} = 13\,\text{TeV}$ with the ATLAS detector. Technical report, ATLAS-CONF-2016-070. Geneva: CERN
5. ATLAS Collaboration (2016) Search for light dijet resonances with the ATLAS detector using a trigger-level analysis in LHC pp collisions at $\sqrt{s} = 13\,\text{TeV}$. Technical report, ATLASCONF-2016-030. Geneva: CERN
6. ATLAS Collaboration (2016) Search for resonances in the mass distribution of jet pairs with one or two jets identified as b-jets with the ATLAS detector with 2015 and 2016 data. Technical report, ATLAS-CONF-2016-060. Geneva: CERN

Printed in the United States
By Bookmasters